Preface

In the annals of human history, there are stories of profound transformation, tales of individuals who, against all odds, underwent remarkable journeys of personal and ideological change. The narrative unfolds within these pages is one story that traverses the vast chasm between denial and enlightenment, skepticism and conviction, apathy, and activism.

The protagonist of this tale, Prof. Jamie Cole, once an ardent climate denier, embarked on a path seldom traveled. Like many in the early 21st century, he once viewed the science behind climate change with skepticism, dismissing the warnings of scientists and experts as mere exaggerations. His beliefs were firmly anchored in a web of misinformation woven over decades, fueled by vested interests and sustained by ignorance. But as the world grappled with the growing impacts of a changing climate - rising temperatures, extreme weather events, vanishing ice sheets, and ecological upheaval - Prof. Cole began to experience a profound awakening. The walls of denial that had once held his convictions crumbled under the weight of mounting evidence and personal experience.

This journey, from denial to acceptance, from indifference to activism, is a testament to the power of knowledge, empathy, and the human capacity for change. It is a story that mirrors our collective struggle with the greatest challenge of our time - climate change. It raises questions about the role of misinformation, the responsibility of individuals in the face of a global crisis, and the potential for redemption and transformation.

Moreover, "A Denier No More" offers readers valuable additional information and resources for those who wish to embark on their own journey toward becoming climate activists. Whether you are an environmental enthusiast looking to deepen your understanding of climate change or someone who has wrestled with doubts about the issue, this book serves as a guidepost for personal growth and transformation.

By reading this book, you will gain valuable insights into the personal and intellectual transformation of a climate change skeptic into a dedicated advocate, and you will discover the tools and resources needed to take meaningful action in the fight against climate change. "A Denier No More" is a captivating narrative that underscores the potential for positive change and the importance of collective action in addressing one of the most pressing challenges of our time.

Acknowledgement

I want to express my deepest gratitude and appreciation to all who have contributed to the creation and completion of my book, "A Denier No More: Leading the Charge for Climate Change." This book has been a labor of love, and I couldn't have accomplished it without the support and assistance of numerous individuals and organizations.

First and foremost, I want to acknowledge the tireless efforts of my family and friends who stood by me throughout this writing journey. Your encouragement, understanding, and patience were invaluable as I delved into the complex and often controversial topic of climate change.

I am deeply grateful to the team at IngramSpark, including the editors, designers, and marketing professionals who played a crucial role in bringing this book to life. Your expertise and commitment to excellence have transformed my manuscript into a polished and professional work.

I would also like to thank the experts and researchers in the field of climate science who generously shared their knowledge and insights with me. Your expertise has enriched the narrative and provided a solid foundation for the story's scientific aspects.

I extend my appreciation to the readers and reviewers who took the time to engage with this book. Your feedback and enthusiasm have been both motivating and enlightening.

Last but not least, I want to acknowledge the Earth itself, the ultimate source of inspiration for "A Denier No More." Our planet's beauty and vulnerability constantly remind us of the importance of addressing climate change.

In writing this book, I hope to contribute to the ongoing conversation about climate change and inspire positive action. Thank you to everyone who has been a part of this journey. Your support has been instrumental in bringing "A Denier No More: Leading the Charge for Climate Change" to fruition.

Disclaimer

This book, "A Denier No More: Leading the Charge for Climate Change," contains a blend of fictional and real-world elements, including names, events, and places. The purpose of these fictional elements is solely for narrative and explanatory purposes. The author wishes to clarify that any resemblance to actual persons is purely coincidental.

The author has taken creative liberties in crafting the story, characters, and settings to enhance the reader's experience and convey the intended themes and messages. While some concepts, historical events, or locations are real and may serve as inspiration, some have been adapted and altered for artistic and storytelling reasons. Readers are encouraged to enjoy the narrative for its educational and entertainment value and not interpret all of it as an accurate representation of historical facts or real individuals.

The author appreciates readers' understanding and suspension of disbelief, allowing them to immerse themselves in "A Denier No More: Leading the Charge for Climate Change." Your journey through these pages is meant to be a captivating exploration of imagination and storytelling.

Thank you for understanding and embarking on this literary adventure with me.

David M Kargbo, Ph.D.
Author

Table of Contents

1

PERSONAL TRANSFORMATION

1.1. THE EARLY DAYS

Jamie Cole, Jr. was born in Sierra Leone. Located on the West African coast, Sierra Leone is a nation with a rich history, diverse culture, and stunning natural beauty. It is rich in natural resources, with a diverse range of minerals, fertile land, and abundant fish stocks. The country shares a border with Liberia and Guinea. It was the good old days. Everything was plentiful in both countries, with a climate conducive to good agriculture and health. Farmers in Sierra Leone and Liberia were rich in crop resources.

Jamie's parents (Ana Cole and Jamie Cole, Sr.) were drug refugees from Columbia who fled a war between the drug cartels and the government and settled first in Honduras and later migrated to Sierra Leone, where Jamie's uncle (Willie Cole, the dad's brother) lived.

Jamie was the second child born to his parents in Sierra Leone. He was a brilliant, thoughtful kid who succeeded in areas that everyone thought impossible. His sister, Marie Cole, was just as bright and was a close confidant of Jamie. Jamie, like Marie, developed a thinking of not necessarily being in the same boat as everyone, even when the idea is popular.

After graduating high school, Jamie attended college in Liberia because he was interested in geology and mining engineering. At that time, Liberia had the best geology and mining engineering schools in the region. Jamie completed his B.Sc. and M.Sc. degrees in geology and decided to stay in Liberia and build a career.

Jamie had his high school sweetheart in Sierra Leone (Musu Conteh) join him in Liberia. Both were engaged to be married. On the eve of their wedding,

Musu discovered the existence of Jamie's son, Alie Cole. When Jamie was in Liberia, he had an affair with a Liberian native (Loucrecia Kollie), with whom he had Alie. Jamie and the boy, Alie, will later become allies when Jamie changes and becomes a climate change advocate.

In a dramatic twist of events, Jamie later discovered that his uncle, Willie, now the head of a political party in Sierra Leone, is his birth father, not Jamie, Sr. Willie had visited Jamie's parents when they were in Honduras, where he had a fling with Jamie's mother. This event will later cause a horrible rift within the family. Upon finding out the truth, Jamie never spoke to his uncle Willie again. Willie died three years later. There were rumors that Jamie had Willie killed. Marie never found out that Jamie was her stepbrother and Willie was Jamie's dad.

1.2.   ADVANCED EDUCATION IN THE USA

Jamie decided to pursue graduate studies in the United States of America (USA), resulting in his family relocating from Liberia to Sierra Leone. With his wife, Musu, pregnant with their second child, Jamie traveled alone to the USA to pursue a Ph.D. in geosciences, specifically in hydro-fracture of geologic formations for efficient oil and natural gas extraction. Musu and Jamie had two children (Laura Cole and Joseph Cole).

Jamie's research was so impressive and successful in securing grant funding that after completing his Ph.D. program, he was hired immediately as an Assistant Professor by a nearby university.

Prof. Cole was on a student visa, which made it difficult for him to get an excellent job in the U.S. To secure employment and stay in the U.S., he married a female friend (Josephine Knudsen) whom he met in graduate school. Prof. Cole never disclosed to his newlywed wife that he was married and had three children back in Liberia. Prof. Cole gradually discontinued communication with his family back in Sierra Leone. He did not even know he was a grandfather when Alie had a child out of wedlock.

1.3.   A PROFOUND CONNECTION

In the vast tapestry of academia, relationships between professors and students often range from distant and impersonal to close and nurturing. It is common for professors, absorbed in their research and the demands of their profession, to interact minimally with students. However, now and then, a unique connection emerges that defies the boundaries of conventional student-teacher relationships. Such is the story of Prof. Cole, who was now a professor of geology, and Carlos, an international student from Honduras studying agricultural science. Their unlikely friendship transcended the traditional roles of mentor and mentee and left a lasting impact on their lives.

Prof. Cole was known throughout the university as a brilliant geologist, admired for his groundbreaking research on hydro-fracture techniques for oil and gas exploration. He was a man of few words, often lost in the intricate world of rocks and minerals. His classes were challenging, and his expectations were high, leaving little room for casual interactions with his students. As a result, many perceived him as unapproachable and distant.

On the other hand, Carlos was an international student filled with enthusiasm and dreams of making a difference in the world, especially in sustainable farming. His journey to the United States was fraught with challenges, from language barriers to cultural differences. He felt like a small fish in a vast ocean of academia.

Prof.Cole and Carlos' paths crossed during a departmental seminar on climate change and its impact on agriculture. Prof. Cole was invited as a guest speaker to share his expertise on how geological factors could affect agricultural practices. Carlos was among the few students who attended the seminar, driven by his thirst for knowledge and desire to connect with experts in his field.

Prof. Cole's presentation was nothing short of captivating. He spoke with authority from years of research and dedication, and his passion for geology was evident in every word he uttered. As Carlos listened, he was struck by the profound connection between geology and agriculture, a link he had never fully appreciated. After the seminar, he mustered the courage to approach Prof. Cole, a man he had long admired from a distance.

Their initial conversation was awkward, to say the least. Prof. Cole, not accustomed to students seeking him for anything other than academic matters, was taken aback by Carlos's enthusiasm. However, he was intrigued by Carlos's genuine curiosity and eagerness to learn. Over time, they began to meet regularly, discussing geology and agriculture and their lives outside of academia.

As their interactions continued, Prof. Cole discovered a warmth and kindness in Carlos that went beyond his passion for agriculture. Carlos shared stories of his homeland, his struggles to adapt to a foreign culture, and his dreams of improving farming practices in Honduras. He was amazed by Prof. Cole's dedication to his field and his commitment to improving the world through his research.

Their friendship deepened as they explored the beauty of their differences. Prof. Cole introduced Carlos to the world of geology, taking him on field trips to study rock formations and mineral deposits. In return, Carlos taught Prof. Cole about sustainable farming practices and the challenges small-scale farmers face in Honduras. They found common ground in their desire to address environmental issues and create a more sustainable future. However, Prof. Cole convinced Carlos that current climate change activities are cyclical and not primarily caused by human activities. Initially, Carlos was skeptical, but over time, he became convinced.

As the years passed, Prof. Cole and Carlos became inseparable friends, their bond transcending the roles of professor and student. They collaborated on research projects that bridged the gap between geology and agriculture, and their work garnered recognition and accolades from their peers. But more than academic success, their friendship brought a profound sense of fulfillment and happiness to both of their lives. This bond will prove extremely useful later when Prof. Cole struggles to embrace the concept of man-induced climate change.

1.4. CLIMATE CHANGE DENIER

Five years later, Prof. Cole became a recognized and renowned researcher and publisher in the field of geosciences. He has also become a prominent climate change denier. He assisted in forming a coalition of climate change deniers and wrote a book that was a best-seller among climate change deniers. Hurricane Andrew hit the U.S. five years before Prof. Cole completed his Doctoral degree program. Most people labeled Hurricane Andrew the result of man-induced climate change, but not Prof. Cole. He believed climate change is not because of man's activities but because of natural cyclical events. In one vital United Nations (U.N.) meeting at the Intergovernmental Panel on Climate Change (IPCC), Prof. Cole presented findings from research conducted at the request of the oil and gas (fossil fuel) industry. He was determined to make the case that human release of greenhouse gases is not the primary driver of climate change impacts.

The IPCC convened the meeting and invited representatives of regions most impacted by climate change. The goal is to collect enough information to help the U.N. decide what developed nations should contribute and compensate poorer nations. These poorer countries are less resilient to the impacts of climate change caused by rising temperatures due to development activities in more affluent countries.

Prof. Cole invited Carlos to join him at the IPCC meeting. Among the people and countries invited to the meeting that Prof. Cole interacted with are Marie Jean-Baptiste (Haiti), Samuel Kamara (Sierra Leone), Dahiru Abakar (Chad), and Maya Rahman (Bangladesh). In all these countries, climate change is an existential threat whose impacts are already causing more significant loss and suffering for the most vulnerable people in their respective countries. Presentations by most climate change experts show that the links between climate change and poverty and climate change and hunger are clear. These presenters showed that many extreme weather events linked to global warming aren't selective regarding a nation's income or gross domestic product (GDP). Indeed, from 2021 to the present, the world saw deadly floods in Europe, record-breaking floods in China, and a combination of extreme winter temperatures and hurricanes in the United States. Historical data shows that many organizations charged with ranking the countries most affected by climate change top their lists with countries like Japan and Germany. Whereas many higher-income countries generally have more resources dedicated to dealing with the unavoidable impacts of climate change, low-income countries do not.

Judging from the reaction of many of the scientists in the meeting, it was clear that Prof. Cole's presentation provided unsubstantiated data that shows that the contribution of the fossil fuel industry to rising temperatures is minimal at best. The fossil fuel industry was so impressed with Prof. Cole that they made him the primary beneficiary of multiple grants – a case of paradox of ignorance.

1.5   SOURCES OF DENIAL

Next, let us examine the potential sources of Prof. Cole's denial of man-induced changes in our climate and its impact on society. Understanding the motives and influences behind skepticism is essential, as they shape the overall narrative and impact efforts to address the climate change crisis.

Financial Benefits

In the annals of academia, the clash of ideas has always been a cornerstone of intellectual growth. Scholars, scientists, and researchers have long engaged in spirited debates to advance knowledge and understanding. However, since time immemorial, money has been a powerful motivator for human behavior. In academia, where research often requires substantial funding, financial incentives can influence the direction, focus, and sometimes even the outcome of the studies. A professor who receives generous grants or sponsorship from industries with vested interests in downplaying the effects of climate change might feel compelled, consciously or unconsciously, to produce research that aligns with these interests. Such financial ties can create a conflict of interest, raising questions about the credibility and integrity of the study. This is the case with Prof. Cole, who, during a time of unsettled science on climate change, chose to be a climate change denier while working for the fossil fuel industry. It is a story that unravels the paradox of ignorance in the face of overwhelming evidence.

To fully appreciate the gravity of this tale, we must journey back to a time when the science surrounding climate change was indeed unsettled, and Prof. Cole was still in Sierra Leone. The late 20th century was marked by scientific discussions, debates, and an emerging consensus that the planet was undergoing unprecedented changes due to human activities. The IPCC diligently compiled data and analysis to present a clearer picture of the impending crisis. It is in the midst of this scientific turmoil that Prof. Cole emerges.

Prof. Cole, now a respected scientist in his field, was a prominent figure in the academic world. He spent decades researching the Earth's climate and natural systems as a geoscientist. His early work was lauded for its rigor and contribution to the field, but as the science of climate change advanced, Prof. Cole found himself at odds with the prevailing consensus. He has become a vocal climate change denier.

Prof. Cole consistently challenged the prevailing consensus on climate change. He questioned the accuracy of climate models, doubted the human role in global warming, and criticized policies to reduce greenhouse gas emissions. For the layperson, his credentials provided a veneer of credibility to his skepticism.

A disconcerting revelation emerged while Prof. Cole's contrarian views gained attention. It came to light that he was on the payroll of major fossil fuel companies, serving as a researcher, a consultant, and an advisor. These industries, often threatened by policies that aim to reduce carbon emissions, have a history of funding climate change denialism to protect their interests. Prof. Cole's affiliation with the fossil fuel industry began to raise questions about the integrity of his research and the motivations behind his climate skepticism.

There is a paradox of ignorance that lies at the heart of Prof Prof. Cole's story. In a world of uncertainty, fostering open debate and rigorous scrutiny of scientific claims is vital to scientific discourse. However, Prof. Cole's denialism was not rooted in a rigorous examination of data or a genuine pursuit of knowledge. Instead, it seemed to be fueled by financial interests. This paradoxical situation, where a scientist, whose duty is to pursue truth, becomes a purveyor of falsehoods, exemplifies the dangers of vested interests in academia.

Political Beliefs

The debate surrounding climate change has long been contentious, with opinions often divided along several lines, one of which is politics. Research has shown that political ideology is a significant predictor of an individual's beliefs about climate change, with conservatives more likely to be skeptical or deny the existence of climate change. This Section explores the case of Prof. Cole, whose skepticism about climate change is influenced by several political factors.

Political Stance: Prof. Cole holds conservative political beliefs that inform his climate change stance. His skepticism is rooted in a broader ideology skeptical of government intervention and regulatory measures often advocated by climate scientists and activists. His position was not merely that the science was uncertain but that there was a deliberate attempt by policymakers to exaggerate the severity of the situation politically. He vehemently downplays the ability of climate models to predict current and future temperatures. Climate models are not perfect but are based on the best available scientific evidence and have been validated through numerous peer-reviewed studies. The consensus among climate scientists is that climate change is real, primarily caused by human activities, and poses a significant threat to our planet.

Prof. Cole's ideology is often associated with a free-market approach prioritizing economic growth over environmental regulations. He raises concerns about the economic impact of taking action to mitigate climate change. Prof. Cole believes that rules and regulations aimed at mitigating climate change would stifle economic growth, infringe on individual liberties, and give excessive power to governmental bodies. He argues that the cost of transitioning to renewable energy sources and implementing other climate mitigation strategies will be detrimental to the economy. Prof. Cole's perspective fails to consider the long-term economic benefits of addressing climate change, such as avoiding the costs associated with extreme weather events, rising sea levels, and other climate-related impacts. Furthermore, the transition to renewable energy will likely create new jobs and stimulate economic growth.

In his lectures and writings, he frequently conflated the science of climate change with policy solutions, arguing against both simultaneously. This conservative position blurs the line between scientific skepticism and policy disagreement, making it difficult for students and peers to discern whether his objections were indeed with the science or the potential governmental responses to the crisis.

The Echo Chamber Effect: The echo chamber effect refers to the phenomenon where people are exposed only to information or opinions that reflect and reinforce their own. Climate deniers are likely to seek out and engage with information that supports their beliefs while avoiding or dismissing information that contradicts them. Prof. Cole was always in the circle of conservative friends and professionals who were climate deniers. This selective exposure to information may have strengthened his denial of man-induced climate change.

The echo chamber effect can also contribute to the polarization of public opinion on climate change. As people are exposed to only one side of the argument, they may become more entrenched in their beliefs and less willing to consider alternative viewpoints. This may have contributed to the strained relationship between Prof. Cole and his wife, as he refused to acknowledge her perspectives.

Within echo chambers, misinformation can spread rapidly and be accepted as truth. This can contribute to the spread of false or misleading information about climate change, further fueling climate denial. Prof. Cole communicates constantly with students. His responses to students may have helped spread incorrect information on climate change and even created new climate deniers.

The echo chamber effect can also inhibit action on climate change. As people are surrounded by others who share their beliefs, they may feel that their actions are justified and that there is no need to take steps to mitigate climate change. For a fiscal conservative like Prof. Cole, this may explain why he argues against pouring resources into addressing climate change.

Impact on Students: One of the more concerning aspects of Prof. Cole's stance was the influence of his climate change stance on impressionable students. When a professor, who is also a climate change denier, teaches students in a way that misrepresents the overwhelming scientific consensus that climate change is real, human-caused, and a serious threat, there could be several negative impacts on the students.

Students could be misinformed about the facts of climate change, including its causes, impacts, and the urgency of addressing it. This misinformation could hinder their ability to make informed decisions and take appropriate actions in response to climate change.

As a climate change denier, Prof. Cole could not have taught his students how to critically evaluate evidence and distinguish between credible and non-credible sources of information. In that case, students are unable to develop the necessary skills to assess other controversial or complex issues critically.

Students in the field of environmental science may lose trust in Prof. Cole and the educational institution if they feel they are not being taught accurate and reliable information. This loss of confidence could have a lasting impact on their educational experience and future learning.

Many students, especially those not majoring in environmental sciences, might not have the tools or knowledge to assess climate change claims critically. For some, Prof. Cole's skepticism may have sown seeds of doubt about the legitimacy of climate science.

As climate change continues to impact the world in various ways, students need to be adequately prepared to address these challenges in their personal and professional lives. If they are not taught the facts about climate change and the importance of mitigating it, they may be ill-equipped to tackle the challenges that arise as a result of a changing climate. They may even become climate change deniers.

Faith

The climate change debate often intertwines with various societal, political, and personal factors that influence individuals' perspectives on the issue. In some cases, religious beliefs significantly shape an individual's perceptions of climate change, as well as their willingness to take action to address the issue.

Many religious traditions emphasize the importance of stewardship of the Earth and its resources and the responsibility to care for creation. For example, the Pope's encyclical "Laudato Si'" calls on Catholics to take action on climate change and care for the planet as a moral responsibility. On the one hand, some religious beliefs can contribute to climate change denialism. This Section will explore the unique position of Prof. Cole, whose skepticism about climate change is also partly rooted in his religious beliefs. His perspective reflects a broader narrative wherein faith and science are often at odds.

At the heart of Prof. Cole's skepticism is the belief in a divine plan that governs the Earth and its climate. He contends that the Earth and its environment are under the protection of a higher power and that humans are not capable of causing significant harm to the planet. Prof. Cole's beliefs also hinge on the idea that the Earth is resilient and self-healing. He argues that the planet has undergone numerous climatic changes throughout its history and has always managed to restore equilibrium. This belief is rooted in the idea of a harmonious and balanced natural world that is capable of regulating itself, with or without human intervention. However, this perspective often neglects the rapidity and severity of the recent changes and the overwhelming data linking them to human actions.

The case of Prof. Cole is not an isolated one. Many people worldwide grapple with reconciling their religious beliefs with scientific evidence. This clash has implications for public policy and education. When influential figures, like professors, propagate views influenced more by personal beliefs than empirical evidence, it can hinder collective efforts to address global challenges.

Psychological Factors

Psychological factors can influence an individual, particularly a professor, to adopt a skeptical stance on climate change despite the overwhelming scientific evidence supporting its existence and impact. Prof. Cole is a learned individual who is usually expected to base opinions on empirical evidence. When such a figure expresses skepticism about climate change, it may seem perplexing, given the wealth of scientific data supporting the phenomenon. The roots of this skepticism can also be traced to a few critical psychological factors. While this author does not claim that Prof. Cole's views directly result from all these factors, it is essential to discuss them to give the reader a complete picture of the potential sources of climate change denial.

Confirmation Bias: Confirmation bias is the tendency to seek, interpret, and remember information that confirms our preexisting beliefs while ignoring or dismissing information that challenges them.

Prof. Cole seeks out information that supports his belief that human activities do not cause climate change. He frequents websites, follows social media accounts, attends conferences, and engages with conservative climate change deniers who share his perspective.

On those few occasions when he was interviewed and presented with information about climate change, Prof. Cole interpreted it in a way that fit his preexisting beliefs. For example, in one interview, he focused on data that appears to show little or no warming while ignoring the overwhelming scientific consensus that climate change is real and caused mainly by human activities. When asked by the interviewer, Prof. Cole remembered information that supported his beliefs more vividly and accurately than information that challenged them. For example, he recalled two cold winters as evidence that the planet is not warming while forgetting about record-breaking heatwaves and other signs of climate change.

Selective picking of data, studies, or articles that support a climate denier's skepticism while ignoring or downplaying the vast body of evidence that contradicts it is a self-reinforcing loop that can solidify the climate denier's suspicion, making him resistant to change.

Psychological Distance: Psychological distance is a cognitive construct that refers to how far away an event or issue feels to a person in terms of time, space, social distance, or hypotheticality. If a climate denier perceives the impacts of climate change as something that will happen in the distant future, they may feel less concerned and motivated to take action.

Similarly, suppose a climate denier believes that the effects of climate change will only impact far-away places and not their local community. In that case, they may not see it as a relevant issue. If a climate denier perceives the science of climate change as uncertain or the solutions as impractical, they may be more likely to dismiss the problem altogether.

Finally, suppose a climate denier does not identify with the people who are most affected by climate change, such as people in developing countries or future generations. In that case, they may feel less empathy and responsibility to address the issue.

It is unclear whether Prof. Cole was affected by any of the above psychological distance factors. Later in this Chapter, when Prof. Cole decided to be a climate crusader, he was clearly moved by the climate impact events in his home country, Sierra Leone, a developing country disproportionately impacted by climate change.

Cognitive Dissonance: Cognitive dissonance is a psychological theory developed by Leon Festinger in 1957. It describes the discomfort people feel when they hold two or more contradictory beliefs, values, or attitudes or when their behavior conflicts with their beliefs or values. In order to reduce this discomfort, people may change their views, acquire new information that supports their ideas, or minimize the importance of the contradiction. For example, a person might value nature and understand the importance of protecting the environment. Still, they might also appreciate economic growth and fear that taking action on climate change will hurt the economy. To resolve the dissonance, they might choose to deny the reality of climate change in order to preserve their belief in the importance of economic growth.

A person might belong to a social or political group that denies climate change. The dissonance between their group identity and the scientific consensus on climate change might lead them to reject the scientific evidence in order to maintain their social identity.

A person might drive a gas-guzzling car or work in an industry that contributes to greenhouse gas emissions. The dissonance between their behavior and the knowledge that such actions contribute to climate change might lead them to deny the reality of climate change in order to justify their behavior.

For a professor who might have invested decades into a particular line of research or belief system, accepting a new or opposing viewpoint, like the anthropogenic influence on climate change, can be unsettling. They might find it easier to dismiss new evidence or findings that challenge their preexisting beliefs rather than reconcile the contradiction.

In Prof. Cole's case, the discord may arise from his long-time research efforts to prove that climate change is not primarily a human-induced phenomenon. Accepting the concept of anthropogenic influence on climate change can be unsettling to Prof. Cole. The professor also belongs to the group of climate change deniers. Ignoring the dissonance between this group identity and the scientific consensus on climate change may be too difficult for Prof. Cole.

Identity-Protective Cognition: People tend to align their beliefs with those of their cultural, social, or professional groups. If a professor identifies strongly with a group that is skeptical about climate change, they might adopt and propagate those beliefs, even in the face of contradictory evidence. Their skepticism might be more about protecting their social identity than a genuine doubt about the science. Identity-Protective Cognition may not have influenced Prof. Cole.

Perceived Threat to Autonomy: Some people resist the idea of climate change because of its implications for policy, regulation, and lifestyle changes.

The idea of being told to change one's way of life, consumption habits, or research focus can be perceived as a threat to personal and academic freedom. Prof. Cole never perceived climate change as a threat to personal and academic freedom.

Fear and Anxiety: Climate change can evoke fear and helplessness, which some individuals may cope with by denying the problem. Acknowledging the enormity of the issue can be psychologically overwhelming. It is worth noting that fear and anxiety can also be used as motivators for action. In the case of climate change, individuals may be more likely to take action to mitigate climate change if they are concerned about the potential impacts on their own lives, their communities, and future generations. Fear and anxiety may not have been psychological factors that influenced Prof. Cole.

The Dunning-Kruger Effect: This cognitive bias refers to the paradox where individuals with low knowledge or expertise in a subject overestimate their abilities, while those with high expertise may underestimate theirs. A professor not specializing in climatology or environmental science might think they understand climate change's nuances better than they do, leading them to draw incorrect conclusions. The Dunning-Kruger Effect may not be a psychological factor that influenced Prof. Cole.

Cultural and Social Influences

A professor's skepticism towards climate change, even when armed with vast academic qualifications, can often be influenced by a melange of cultural and social factors. Like the psychological factors, this author does not claim that Prof. Cole's views directly result from all these cultural and social factors. Nevertheless, it is essential to discuss them to give the reader a more complete picture of the sources of climate change denial.

Economic Interests and Livelihoods: Understanding the social and economic milieu from which Prof. Cole originates is essential. In regions that heavily depend on fossil fuel industries (such as the U.S.), there's often a communal bias against accepting the detrimental effects of climate change. If a professor's personal or familial livelihood connects to these industries, their skepticism might be influenced by the perceived economic threat of widespread acceptance of climate change realities. This is the case with Prof. Cole, which is covered in more detail in the Financial Benefits Section of the Chapter. When the science on climate change was not as settled as it is now, Prof. Cole, instead of working diligently to provide solutions to the climate change problem, chose to be a climate denier while at the same time working for the fossil fuel industry.

Media Consumption: The media environment in which a person is immersed plays a significant role in shaping their beliefs. Some media outlets, driven by political or economic motives, amplify the voices of climate change skeptics, thereby influencing public and academic opinions. For example, suppose the media does not focus on the scientific consensus of climate change as a human-induced problem, the urgency of action, or the economic costs of inaction. In that case, it may even create more climate change deniers. Prof. Cole consistently consumes conservative media that downplays or denies climate change. His perspective is likely to be shaped by that narrative. For more detailed information on the role of the media in climate communication, please refer to Chapter 10 (Climate Communication).

Peer Influence and Echo Chambers: As previously noted in the Political Beliefs Subsection of this Chapter, academia is not immune to the influence of echo chambers. Like anyone else, professors can be surrounded by peers who reinforce their skepticism. If one's close academic and social circles predominantly question or reject the consensus on climate change, this can create a feedback loop that solidifies and deepens one's skepticism. Prof. Cole's social circle peers could have influenced him.

Education and Pedagogy: Lastly, how climate science is taught and introduced to individuals can influence their perspectives. If, during their formative years, the professor was introduced to climate science in a manner that seeded doubts, or if they were exposed to instructors who themselves were skeptics, this early education could have a lasting impact on their beliefs. This concept may have influenced Prof. Cole's students. Its effect on students is covered in more detail in the Political Beliefs Subsection.

Cultural Identity and Tribalism: In many societies, beliefs about climate change have been entangled with cultural identities. If the professor hails from a culture or subculture with normative skepticism towards climate change, this background could influence their opinions. For instance, in certain conservative circles in the U.S., rejecting the mainstream narrative about climate change has become a badge of cultural identity. For individuals deeply embedded in these cultures, accepting the consensus on climate change might be perceived as a betrayal of one's cultural identity. Prof. Cole hails from an African culture, and his views on climate may not have been a result of his cultural identity and tribalism.

Reaction to Perceived Alarmism: Some individuals develop skepticism as a counter-reaction to what they perceive as alarmist rhetoric. Suppose a professor believes that certain groups exaggerate the consequences of climate change for political or other reasons. In that case, he might distance himself from the mainstream viewpoint, even if he accepts the basic science. Prof. Cole did not initially belong to this group as he rejected basic science.

1.6 METAMORPHOSIS OF A CLIMATE DENIER

The consequences of Prof. Cole's actions were profound. In a time when swift and decisive action on climate change was imperative, his vocal denialism contributed to a sense of uncertainty and confusion among the public and policymakers. The fossil fuel industry used his arguments to sow doubt about the reality of climate change, thereby delaying crucial actions to mitigate its effects.

The story of Prof. Cole, a climate change denier working for the fossil fuel industry during a time of unsettled science, is a cautionary tale. It highlights the ethical dilemmas that can arise when personal interests clash with scientific responsibility. Ultimately, it is a reminder that the pursuit of knowledge must always be guided by integrity, honesty, and a commitment to the betterment of humanity, especially in matters as critical as climate change.

However, Prof. Cole went from a climate denier and industry advocate to a climate crusader over time. In the annals of climate change history, few stories are as riveting as that of Prof. Cole, once an unyielding climate change denier working hand-in-hand with the fossil fuel industry. Today, he stands as a symbol of hope and transformation, having crossed the divide to become a fervent advocate for urgent action on climate change. This remarkable journey from skepticism to activism underscores the power of science, conviction, and the capacity for change, even in the most unexpected places.

Prof Cole's transformation from climate change denier to climate advocate was not sudden but gradual, marked by several key events and realizations.

Climate Refugees

The year was 2016. Prof. Cole learned that Carlos's parents are now climate change refugees. They lost their land, crops, and cattle due to extreme weather events, which included hurricanes and droughts, exacerbated by climate change. These losses destroyed their livelihood, forcing them to leave their homes for better opportunities. When the Rodriguez family arrived at the U.S. southern border, they faced the challenge of gaining entry into the U.S., which various immigration policies have complicated. Penniless, they called their son Carlos for help. Prof. Cole was so devastated to hear the story that he offered to help.

One year later (2017), Prof. Cole faced another climate-induced disaster. Prof. Cole had been out of his homeland, Sierra Leone, for over ten years, during which time he had little or no communication with anyone, not his wife, children, or parents. Then came the record mudslide, a devastating natural disaster, on August 14 in the Regent area near the country's capital, Freetown. A portion of Sugar Loaf Mountain collapsed after heavy rainfall, causing a massive mudslide and flooding that engulfed homes and communities. Factors such as heavy rainfalls, deforestation, unregulated construction, and inadequate infrastructure contributed to the severity of the disaster. The mudslide exposed the climate-induced vulnerabilities and challenges faced by Sierra Leoneans. The disaster resulted in the death of over 1,000 people, with hundreds more missing and thousands displaced from their homes.

One of those displaced families was Prof. Cole's family. A year later, in 2018, Prof. Cole read about the disaster and that his family was one of those displaced. Prof. Cole immediately embarked on a journey to his homeland, which he had not visited for over ten years. On arrival, he found his family (Jamie Sr., Ana, and Marie) in a camp for displaced persons. Even more shocking, Prof. Cole's son, Alie, was at the camp with his wife. Alie and his family are also climate refugees. Prof. Cole was shocked when he found that he had a granddaughter, Ana Cole, Jr. (named after her grandmother, Ana Cole). Ana Cole Jr. had been forced to marry because of her parent's financial difficulties. Prof. Cole had a lengthy discussion with his sister on whether refugee status results from climate change. Like Prof. Cole, Marie is an ardent climate change denier. She doesn't believe that man's activities are primarily responsible for the change in climate. Prof. Cole helped the family and moved them from the camp to a large, comfortable home in Freetown. He promised to take care of them, and so he did.

A Daughter's Sacrifice

Ana, Prof. Cole's granddaughter, whom he knew nothing about, is a young woman whose life is turned upside down by the impacts of climate change. As her community in the Northern District of Sierra Leone faces climate change disasters, her family is forced to flee their home and become refugees at the same camp that houses her grandmother and grandfather displaced due to the mudslide in Freetown. Struggling to adapt to their new circumstances and facing extreme financial hardship, Alie (Ana's dad) feels pressured to marry her off to secure her future and alleviate their economic burden. Ana longed for a future that seemed within reach until the rains in Sierra Leone dwindled and the soil had lost fertility. Her dreams of becoming a nurse, to care for her people in this time of need seemed to fade with every passing day. Despite her reluctance and desire for autonomy, Ana is forced into a marriage she does not want. She agreed to sacrifice herself for her family. She was only 15. Prof. Cole was determined to end the marriage, but Ana was with a child. This episode was devastating to Prof. Cole.

A Stormy Climate

Prof. Cole was known throughout his university as a contrarian. But it wasn't just his colleagues and students who challenged his views. His wife Josephine was his most prominent critic. She bore him a daughter (Soria).

Josephine had always been passionate about the environment. She had met Prof. Cole at a climate march, where she had been protesting, and he had been observing. Despite their differing views, they were drawn to each other's passion and intelligence and soon found themselves in a whirlwind romance.

As the years passed, however, their differing views on climate change became a point of contention. Josephine often found herself frustrated by Prof. Cole's refusal to acknowledge the overwhelming scientific consensus on climate change. She couldn't understand how someone so intelligent and logical could deny something happening before their eyes. Prof. Cole never revealed his relationship with the fossil fuel companies to her. Josephine suspected the existence of the relationship but never discussed it with Prof. Cole.

Prof. Cole felt that Josephine was being swept up in hysteria. He frequently explained to her that the climate was constantly changing and that human activity was not the primary driver of these changes. He saw himself as a defender of reason and scientific inquiry, and he was frustrated by what he saw as Josephine's unwillingness to consider alternative viewpoints.

One summer evening, the tension reached a boiling point as they sat on their porch watching a storm roll in. The air was thick with humidity, and the dark clouds seemed to mirror the storm brewing between them.

"I just don't understand how you can be so blind to the evidence," Josephine said, her voice rising over the rumbling thunder.

"And I don't understand how you can be so blind to the possibility that you might be wrong," Prof. Cole replied his tone just as heated.

The argument escalated until they both shouted, their words lost in the storm's roar. Eventually, exhausted and soaked to the bone, they retreated inside, each reverting to their respective corners of the house.

In the following days, the tension between them remained thick and unspoken. They moved around each other like two planets in orbit, close but never quite meeting.

But as the days turned into weeks, Prof. Cole began reflecting on the toll their disagreements took on their marriage. He realized he had been stubborn and close-minded, refusing to consider Josephine's perspective. He started researching, diving more into the scientific literature on climate change – more than ever before.

The more he read, the more he began seeing the validity of Josephine's viewpoint. The evidence was overwhelming, and he couldn't understand how he had been so blind to it before. One evening, as they sat in their living room, Prof. Cole turned to Josephine and said, "You were right honey. I've been a fool."

Josephine looked at him in surprise. "Really?"

"Yes," Prof. Cole replied, taking her hand. "I'm sorry for being so stubborn.

I love you and don't want our differing views on this issue to come between us."

Josephine smiled, tears welling up in her eyes. "I love you too. And I'm sorry for not being more understanding of your perspective."

But it was too late. For more than a year, Prof. Cole and Josephine slept in separate rooms. Josephine had started drinking heavily. Under the influence of alcohol, Josephine had several extramarital affairs. Prof. Cole finally discovered the extramarital romances. Josephine confessed to Prof. Cole and asked for a divorce. Prof. Cole was devastated, but he granted her the divorce.

The Change of Heart Lecture

A pivotal moment for Prof. Cole was when he attended a scientific lecture on climate change and was exposed to the latest research findings. One day, a renowned climate scientist, Dr. Eleanor Davies, widely respected in the scientific community, was invited to lecture at Prof. Cole's university. The university organized the lecture for students and the public who may not be versed in the science of climate change but are interested in what is happening to our planet Earth.

Dr. Davies explained in simple terms, with graphics illustrations, that humans are changing the climate by releasing a lot of extra greenhouse gases into the atmosphere. Greenhouse gases, like carbon dioxide (CO_2) and methane (CH_4), act like a blanket around the Earth, trapping heat from the sun and warming the planet. Here's how humans contribute to the warming of planet Earth.

Burning Fossil Fuels: We use fossil fuels like coal, oil, and natural gas for things like driving cars, generating electricity, and heating our homes. When we burn these fuels, they release CO_2 into the air.

Deforestation: Cutting down trees, especially in large forests, reduces the planet's ability to absorb CO_2 from the atmosphere. Trees act like Earth's natural air filters, so removing them means more CO_2 stays in the air.

Agriculture: Some farming practices (e.g., raising cattle) produce methane, a potent greenhouse gas. Additionally, using certain fertilizers can release nitrous oxide, another greenhouse gas.

Industrial Processes: Certain industrial activities release greenhouse gases into the atmosphere. For example, the production of cement generates lots of CO_2.

Waste: When we throw away food and other organic materials, they break down in landfills and produce methane.

All these extra greenhouse gases are causing the Earth's average temperature to rise, leading to global warming. This warming can have many adverse effects, including more extreme weather, melting ice caps, rising sea levels, and disruptions to ecosystems. Chapter 2 (Climate Science) provides more detailed information on the causes and impacts of climate change.

Dr. Davies explained her research on melting polar ice caps and its impact on sea levels. Prof. Cole attended the lecture, fully prepared to dispute her findings. However, as Dr. Davies presented her research, something began to change in Prof. Cole. Dr. Davies showed data from multiple studies that all pointed to the same conclusion: the melting of the polar ice caps was accelerating, and human-induced climate change was a significant factor. She also shared heartbreaking images of polar bears struggling to find food and island communities threatened by rising sea levels. Dr. Davies spoke with a passion and conviction that was hard to ignore.

Prof. Cole left the lecture deep in thought. He thought about his divorce, which did not have to happen, and the forced marriage of his granddaughter, who became a climate refugee. Now, Prof. Cole is confronted with undeniable evidence against everything he believes. He decided to do something he had never done before; conduct advanced climate change research and delve more into climate change with an open mind.

Witness Convert

Finally, an unexpected trip he took with his graduate students and Carlos to the Arctic Circle for research was a life-altering experience. He witnessed firsthand the rapid melting of glaciers, the shrinking polar ice caps, and the plight of polar bears struggling for survival. Right at that moment, he thought about people dear to him who became climate refugees (his parents, sister, son, granddaughter, Carlos' parents), his failed marriage, and his granddaughter's forced marriage. He thought about the lecture by the highly respected Dr. Davies. Prof. Cole realized he could no longer ignore the undeniable evidence. In fact, he felt he may have contributed to the problem by his long-time denial of climate change.

Returning to his home, he continued to see the devastating effects of climate change in vulnerable communities, including increased wildfires, extreme weather events, and coastal erosion. These and the other experiences left him deeply troubled as he continued to understand the human suffering and ecological damage caused by climate change.

For months, Prof. Cole immersed himself in the latest studies and spoke with respected climate scientists worldwide. He analyzed data on rising temperatures, melting ice caps, and increasing extreme weather events. The more he learned, the more he realized how wrong he had been.

Prof. Cole now faces a profound internal conflict. He knew that he could not remain a climate denier any longer, nor could he continue working for an industry that contributed significantly to the crisis. His denial had cost him personally a lot. It was a time of reckoning. He reevaluated his values and questioned the ethics of the fossil fuel industry's practices.

Prof. Cole was finally ready to admit the existence of man-induced climate change and its impact. He penned an op-ed for a prominent science journal, detailing his journey from climate denier to climate advocate. He acknowledged the overwhelming consensus among scientists that human activities contributed to climate change and urged others to act.

Prof. Cole's admission sent shockwaves throughout the academic community. Many were surprised by his change of heart but welcomed him with open arms. Prof. Cole became an advocate for climate action, using his platform to educate others and encourage them to reduce their carbon footprint.

Prof. Cole's story is a testament to the power of open-mindedness and the importance of seeking the truth, no matter where it leads.

The Climate Crusader

Prof. Cole has now emerged as a leading voice in the fight against climate change. He was no longer content with denying the problem; instead, he wanted to be part of the solution. Prof. Cole began to speak out about the urgency of climate change, using his credibility as a former researcher and industry insider to reach a broader audience.

Efforts to Address Climate Change: Now committed to efforts to address climate change, Prof. Cole is actively involved in various initiatives, including the following research projects and advocacy campaigns to mitigate the impact of climate change.

Prof. Cole revamps the curriculum in his department to include comprehensive courses on climate science, renewable energy, and sustainable practices. He also organizes workshops and seminars with experts in climate science to educate students and colleagues on the reality and severity of climate change. The professor leads a team of scientists researching innovative solutions to reduce greenhouse gas emissions. He is also involved in studying the impact of climate change on different ecosystems and exploring ways to protect biodiversity. He has begun to develop or improve innovative technologies that can help reduce our dependence on fossil fuels and transition to cleaner energy sources. He is also involved in collaborative research projects with other universities and research institutions worldwide to find global solutions to climate issues. Prof. Cole became one of the first researchers to explore ways to use artificial intelligence and machine learning to understand better and mitigate the impacts of climate change.

At the international level, Prof. Cole is a vocal advocate for strong climate policies and regulations, regularly meeting with policymakers to push for stricter environmental laws. He also writes articles and opinion pieces in prominent journals and magazines to influence public opinion and policy decisions related to climate change. As a member of several international climate panels and forums, Prof. Cole works with global leaders and scientists to develop strategies to combat climate change.

At the community level, Prof. Cole actively engages with local communities to spread awareness about climate change and its effects. He formed an NGO to raise funds that will support other NGOs. His NGO partnered with a local NGO to initiate tree-planting drives and clean-up campaigns that involve the community in direct action against climate change.

Prof. Jamie Cole has now dedicated his professional and personal life to addressing climate change on multiple fronts. By combining education, research, community engagement, global collaboration, advocacy, and innovative solutions, Prof. Cole plays a crucial role in the worldwide fight against climate change.

Chapter 3 (Grassroots Activism) provides a more complete picture of grassroots activism to combat climate change.

Advocate for Environmental and Climate Justice: Prof. Cole has dedicated himself to advocating for environmental and climate justice. He utilized his academic platform to educate students and colleagues about the realities of climate change and the need for urgent action. He collaborated with other professors to develop and teach courses that focus on the intersection of climate science, social justice, and public policy. As an ardent researcher, Prof. Cole collaborated with other colleagues to research efforts toward understanding climate change's social and economic impacts on vulnerable populations. Prof. Cole uses his research findings to advocate for policies that address the root causes of environmental injustice and promote equitable solutions to climate change. He leverages his academic credibility to challenge climate denialism and promote informed public discourse on climate change.

Prof. Cole worked closely with local communities, especially those disproportionately affected by climate change and environmental degradation. He collaborated with grassroots organizations and activists to amplify the voices and concerns of marginalized communities in the climate justice movement. Prof. Cole occasionally speaks at conferences, community events, and on media platforms to raise awareness about climate justice issues.

Finally, Prof. Cole actively engages with local, national, and international policymakers to advocate for strong climate policies prioritizing social and environmental justice. He also provides expert testimony and scientific evidence to inform policy decisions and hold governments accountable for their climate commitments.

By adopting a holistic approach that combines education, community engagement, research, advocacy, and policy collaboration, Professor Cole has become a prominent figure in fighting environmental and climate justice, working tirelessly to ensure a just transition to a sustainable and equitable future. Chapter 8 (Climate Justice) provides additional details on Climate Justice.

Advocate for the Rights and Well-being of Climate Refugees: Prof. Jamie Cole, who once denied the reality of climate change, finds himself as one impacted by climate change. His family members have been climate refugees. He was very excited to engage in activities related to climate refugees.

First, the professor felt it necessary to stay informed about developments related to climate change and refugees. This allows him to respond quickly to emerging issues and opportunities for advocacy. He uses his platform as a professor to educate students and the public about the reality of climate change, its impacts, and the plight of climate refugees. He gives lectures, writes articles, and engages with the media on issues related to climate refugees. He supports research on climate refugees, including their needs, their challenges, and potential solutions to support them. His research results have been used by many to inform policy and humanitarian efforts. Prof. Cole organizes workshops, seminars, and information sessions to educate others about climate refugees and ways to help. He encourages Carlos, who has artistic skills, to create creative projects that raise awareness about climate refugees and their stories. Carlos' parents were also climate refugees.

Once a conservative, Prof. Cole now votes for leaders prioritizing climate change mitigation and refugee rights. Prof. Cole lobbies for policies that protect the rights of climate refugees. This includes advocating for refugee status for climate migrants, advocating for more robust climate policies, and pushing for humanitarian aid to affected regions. On several occasions, he has contacted elected officials at the local and national levels to express his concerns about climate refugees and ask them to support policies that address this issue. With the help of Carlos, Prof. Cole starts conversations about climate refugees with his friends, family, and colleagues. He encourages open discussions to increase awareness and empathy for the issue.

Prof. Cole joins and occasionally organizes peaceful protests and demonstrations to draw attention to the plight of climate refugees. These events helped pressure governments and organizations to take action. Climate change is a global issue, so Prof. Cole connects with advocates and organizations worldwide to share strategies and experiences. On a national level, he advocates for sustainable practices and policies that can mitigate the effects of climate change. He argues that this will indirectly reduce the number of climate refugees in the future.

Prof. Cole takes personal action to support climate refugees by volunteering, donating to relevant organizations, and using his lifestyle and choices to reduce his carbon footprint. His NGO raised significant funds that he used to assist other NGOs and advocacy groups that are actively working to support climate refugees.

By taking these actions, Prof. Cole, a former climate change denier, now significantly contributes to the fight against climate change and the support of those most affected by it, including climate refugees.

Chapter 9 (Climate Refugees) provides additional information on climate refugees.

Transition to Renewable Energy, Sustainability & Conservation: Prof. Cole utilized his extensive knowledge to educate the public about the urgency of the crisis and the imperative need for sustainable solutions. Chapter 5 (Renewable Energy, Sustainability & Conservation) details the transition to renewable energy and how to address the environmental, social, and economic challenges our dependence on fossil fuels poses.

Prof. Cole's journey from skepticism to advocacy is an inspiring testament to the power of personal transformation. His story is a powerful reminder that change is possible and essential in the face of the climate crisis. His journey from a climate change denier deeply entrenched in the fossil fuel industry to a passionate advocate for climate action is a testament to the transformative power of science, conscience, and the human spirit. Prof. Cole's story is a beacon of hope, illustrating that anyone, regardless of their past beliefs or affiliations, can become an agent of change in the critical battle against climate change. In a world where the stakes are higher than ever, we must look to examples like Prof. Cole to inspire us to take action before it's too late. Nevertheless, we should also be aware that many climate change deniers face psychological barriers that prevent them from accepting the reality of climate change. Their transition from a climate change denier to a climate change crusader will face many challenges.

1.7 TRANSFORMATIONAL CHALLENGES AND SOLUTIONS

Like many individuals who undergo a profound change in their perspective on a controversial issue, Prof. Cole faced several significant challenges when transitioning from a climate change denier to a climate change crusader.

Cognitive Dissonance: Prof. Cole had to confront the cognitive dissonance of changing one's beliefs. This process was challenging, as it involved admitting that one was wrong in one's previous stance. For Prof. Cole, this meant reevaluating the scientific evidence he had previously dismissed or interpreted differently. It required a willingness to be open-minded and to accept that his understanding of climate science had been flawed. This can be a humbling and challenging process, especially for someone in an academic position who is expected to be an authority on the subject matter.

Prof. Cole's new line of research and open-mindedness means he does not harbor contradictory beliefs. Therefore, he experienced no psychological discomfort in his new role as a climate crusader. He found it easier to thoroughly investigate and even accept new evidence or findings that challenged his preexisting beliefs.

Credibility: The climate denier turned crusader faced the challenge of re-establishing his credibility. As a climate change denier, he had published articles and given talks inconsistent with the scientific consensus on climate change. To become a climate crusader, he would need to acknowledge his past mistakes and work to correct them. This involved retracting or revising previous publications and publicly acknowledging his change in perspective. Prof. Cole engaged in the process of rescinding and sometimes revising previous publications and publicly acknowledging his shift in perspective. This process was time-consuming and was not always successful. Some of his peers and

31

others were unwilling to forgive or forget his previous stance.

Financial: One significant step in Prof Cole's transformation was discontinuing his research for the fossil fuel industry. This decision was difficult, as it meant giving up lucrative research grants and financial incentives. However, his newfound commitment to climate action outweighed these considerations. He downsized everything from a luxury home in the countryside to a three- bedroom apartment in a run-down neighborhood.

Confirmation Bias: To overcome confirmation bias, Prof. Cole designed communication strategies that encourage critical thinking and open-mindedness. He sought out multiple sources of information and considered various perspectives. By fostering a culture of intellectual curiosity and humility, Prof. Cole became more receptive to climate change data and overcame his previous confirmation bias.

Psychological Distance: To overcome the psychological distance barrier, Prof. Cole carefully observed climate change's local and immediate impacts that are relevant to people's lives. In one presentation to government officials, he explained how current weather patterns and changes in freshwater availability have directly affected their community's well-being. Connecting climate change to concrete, relatable consequences helped Prof. Cole reduce the psychological distance.

Political and Cultural Identity: By framing climate change as a bipartisan or non-partisan issue, Prof. Cole overcame the political and cultural identity barrier. At the university, he invited leaders and organizations from diverse backgrounds that support climate action as guest lecturers. His presentations inside and outside the classroom also emphasized the moral and ethical aspects of addressing climate change, appealing to shared values that transcend political or cultural differences.

Discounting the Future: Prof. Cole emphasized early climate action's economic and societal benefits in interacting with students and the public. He showcased examples of sustainable practices that offer short-term benefits, such as cost savings, and long-term advantages, such as environmental preservation. By demonstrating that addressing climate change can align with immediate self-interest, Prof. Cole motivated individuals to act.

Social Norms and Conformity: To overcome this barrier, Prof. Cole gradually avoided climate-denier social groups. Instead, he promoted and interacted frequently with individuals and communities that have embraced sustainable practices and received positive recognition. He encouraged open conversations about climate change within social networks, creating dialogue and information-sharing spaces.

1.8 SUMMARY

In summary, the case of Prof. Cole serves as a reminder of the complex web of factors that can influence one's beliefs. A brilliant African from a conservative family became a well-respected researcher who, during a time of unsettled science on climate change, chose to be a climate change denier while working for the fossil fuel industry. This story unravels the paradox of ignorance in the face of overwhelming evidence.

From confronting his cognitive dissonance to facing potential backlash from peers and the public, re-establishing his credibility, and becoming an effective advocate for climate action, Prof. Cole had to navigate a complex landscape of personal and professional hurdles. His journey is a testament to the power of an open mind and the willingness to change in the face of overwhelming evidence. Still, it also highlights the difficulties that can arise when one changes their stance on a contentious issue.

Navigating the challenges of becoming an effective advocate for climate action requires a deep understanding of the scientific evidence supporting the reality of climate change and the ability to communicate this information to various audiences. It also required the development of strategies to engage with and persuade those who may still be skeptical of climate change.

The climate change debate underscores the importance of ensuring that policy disagreements do not overshadow or distort the underlying science. After all, the planet's future hinges not on political ideologies but on our collective response to the genuine challenges posed by a changing climate.

The intersection of faith and science in the climate change debate is a complex and multifaceted issue. While Prof. Cole's religious beliefs give him a unique perspective on climate change, they also highlight the potential for conflict between faith and empirical evidence. It is essential to foster open and respectful dialogue between religious and scientific communities to bridge the divide and work collaboratively toward addressing the pressing challenge of climate change.

The academic community primarily stands behind the consensus on climate change. Still, it's essential to understand the factors influencing skepticism within its ranks. By dissecting the cultural and social influences that shape a professor's beliefs, we can foster more productive conversations and perhaps bridge the divide in understanding this global challenge.

Finally, Prof. Cole's skepticism regarding climate change may not be reduced to mere ignorance or a lack of understanding of the evidence. Some or all of the psychological factors play a significant role in shaping beliefs, and these can act as powerful barriers to accepting the consensus on climate change. Recognizing and addressing these factors is crucial if we hope to bridge the divide and foster a more informed, collaborative approach to tackling the challenges of a changing climate. As an advocate for climate action, one must address these barriers by tailoring one's approach to engage individuals' cognitive, emotional, and social dimensions. By fostering a sense of urgency, relevance, and unity, one could help more people accept the reality of climate change and take meaningful steps to address it.

2

CLIMATE SCIENCE

Climate change is a complex and well-established scientific phenomenon supported by an overwhelming consensus among climate scientists. This Chapter provides key concepts and evidence that helped convince Prof. Cole about the human footprints of climate change.

2.1 THE GREENHOUSE EFFECT

Greenhouse gases (GHGs) are gases in Earth's atmosphere that can trap heat from the sun, leading to a warming effect on the planet's surface. This natural phenomenon, the greenhouse effect, is essential for maintaining Earth's temperature within a habitable range. Without some greenhouse gases in the atmosphere, Earth would be too cold to support life as we know it.

The primary greenhouse gases in Earth's atmosphere include:

Carbon Dioxide (CO_2): Carbon dioxide is the most well-known greenhouse gas and is a significant contributor to the greenhouse effect. It is released into the atmosphere by burning fossil fuels (coal, oil, and natural gas), deforestation, and other industrial processes.

Methane (CH_4): Methane is another potent greenhouse gas, with a much higher heat-trapping capacity per molecule than CO_2. It is released from sources like livestock digestion, rice paddies, landfills, and the production and transport of fossil fuels (natural gas and oil).

Nitrous Oxide (N_2O): Nitrous oxide is produced by agricultural and industrial activities and combustion processes. It has a significant warming potential and contributes to the greenhouse effect and stratospheric ozone depletion.

Water Vapor (H_2O): Water vapor is the most abundant greenhouse gas in the atmosphere, but its concentration is primarily controlled by temperature and natural processes. Human activities do not directly emit significant amounts of water vapor.

Ozone (O_3): Ozone in the lower atmosphere (troposphere) is considered a greenhouse gas, although it is also an essential component of the ozone layer in the upper atmosphere (stratosphere). Ground-level ozone is produced by chemical reactions involving vehicle pollutants, industrial processes, and other sources.

Fluorinated Gases: Certain synthetic compounds, such as hydrofluorocarbons (HFCs), perfluorocarbons (PFCs), and sulfur hexafluoride (SF_6), are used in various industrial applications, including refrigeration, air conditioning, and electronics manufacturing. These gases are potent greenhouse gases with long atmospheric lifetimes.

As previously noted, these greenhouse gases trap heat from the sun by a process known as the greenhouse effect. The sun emits energy through visible light and other electromagnetic radiation (Fig. 2.1).

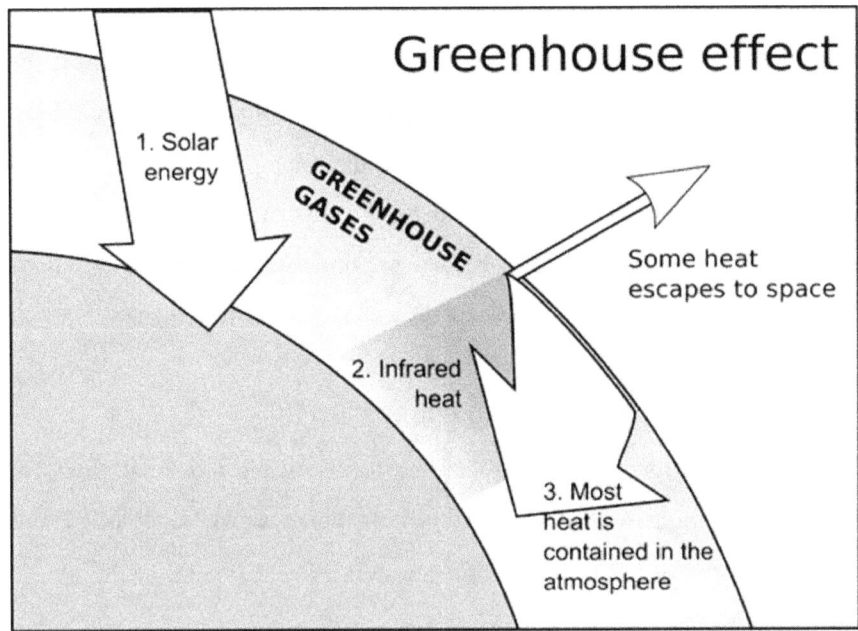

Fig. 2.1: The greenhouse effect. *Credit:* Natural Energy Hub
https://naturalenergyhub.com/environmental-hazards/greenhouse-effect-causes-outcome/ (Accessed November 10, 2023)

This energy (called incoming solar radiation) travels through space and reaches the Earth's surface, where it is absorbed and warms the surface. The Earth then emits some of this absorbed energy in the form of infrared (IR) radiation, which has longer wavelengths than the incoming solar radiation. The greenhouse gases in the Earth's atmosphere, such as CO_2, CH_4, and H_2O, have molecules with the ability to absorb and re-emit infrared radiation. These gases are transparent to incoming solar radiation but absorb and trap some outgoing infrared radiation. When greenhouse gases absorb the infrared radiation emitted by the Earth's surface, they become energized. They then re-emit this energy in all directions, including back toward the Earth's surface. This absorption and re-emission process effectively slows the escape of heat energy from the Earth into space. It's like a blanket around the planet, trapping heat and keeping the

Earth's surface warmer than if these gases were absent.

Human activities, such as burning fossil fuels (coal, oil, and natural gas) and deforestation, have increased the concentration of greenhouse gases in the atmosphere, mainly CO_2. This enhanced greenhouse effect is leading to global warming and climate change, as it intensifies the trapping of heat, causing the Earth's average temperature to rise, with potentially harmful consequences for ecosystems and human societies. Efforts are underway worldwide to reduce greenhouse gas emissions and mitigate the impacts of climate change.

2.2   HUMAN ACTIVITIES AND INCREASED CO_2

The evidence that burning fossil fuels, deforestation, and industrial processes release vast amounts of CO_2 into the atmosphere is well-established and supported by a vast body of scientific research. Here are some of the critical pieces of evidence.

Measuring Atmospheric CO_2 Concentrations

Atmospheric CO_2 concentrations have been measured directly since the late 1950s, most notably at the Mauna Loa Observatory in Hawaii. These measurements show a clear increase in CO_2 levels over time, from about 315 parts per million (ppm) in 1959 to over 410 ppm in recent years (Fig. 2.2).

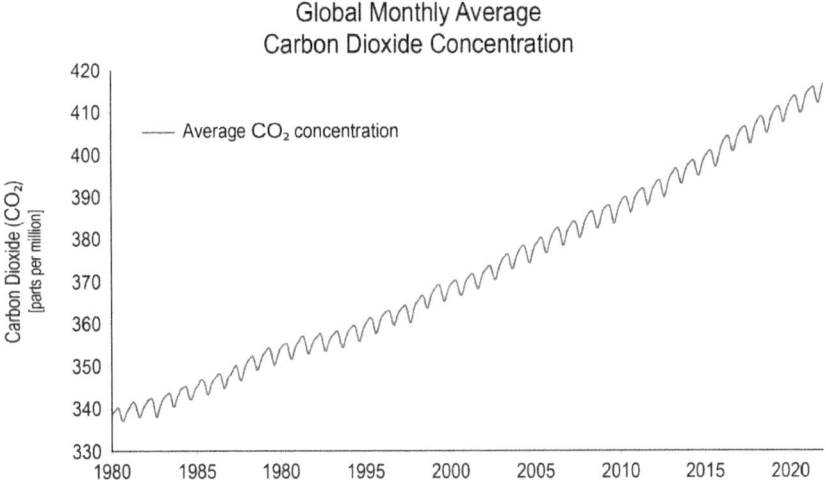

Fig. 2.2: Graph depicting the upward trajectory of carbon dioxide (CO2) in the atmosphere as measured at the Mauna Loa Atmospheric Baseline Observatory. *Credit:* NOAA and Scripps Institution of Oceanography. https://nypost.com/2021/06/08/carbon-dioxide-levels-hit-50-higher-than-preindustrial-time/amp/ (Accessed November 10, 2023)

Emission Inventories

Comprehensive inventories of global greenhouse gas emissions, such as those compiled by the Intergovernmental Panel on Climate Change (IPCC), detail the sources of CO_2 emissions. These inventories show that burning coal, oil, and natural gas is the largest source of anthropogenic (human-caused) CO_2 emissions.

Deforestation and other land-use changes are also significant sources of CO_2 emissions. Trees absorb CO_2 from the atmosphere, but when they are cut down or burned, that stored CO_2 is released into the atmosphere.

Carbon Isotope Analysis

The isotopic composition of atmospheric CO_2 provides a fingerprint that can distinguish between CO_2 from natural sources and CO_2 from burning fossil fuels. Carbon from fossil fuels has a different isotopic signature than carbon from recent biological sources, and analysis of this signature provides further evidence that the increase in atmospheric CO_2 is primarily due to human activities.

Satellite Observations

Satellite data have provided another way to measure CO_2 concentrations in the atmosphere and observe how they change over time and space (NASA, 2023). For example, NASA's Orbiting Carbon Observatory 2 and 3 (OCO-2 & OCO-3) missions have provided detailed measurements of CO_2 concentrations around the globe, providing further evidence that the increase in atmospheric CO_2 is primarily due to human activities.

The consensus among climate scientists is that the increase in atmospheric CO_2 concentrations since the Industrial Revolution is primarily due to burning fossil fuels, deforestation, and industrial processes. This increase in CO_2 is a major driver of global climate change.

2.3   RISING GLOBAL TEMPERATURES

Like the greenhouse gases, the Earth's average surface temperature has steadily risen over the last century. According to data reported by the National Oceanic and Atmospheric Administration (NOAA), the Earth's temperature has risen by an average of 0.14 °F (0.08 °C) per decade since 1880, or about 2.0 °F (Fig. 2.3).

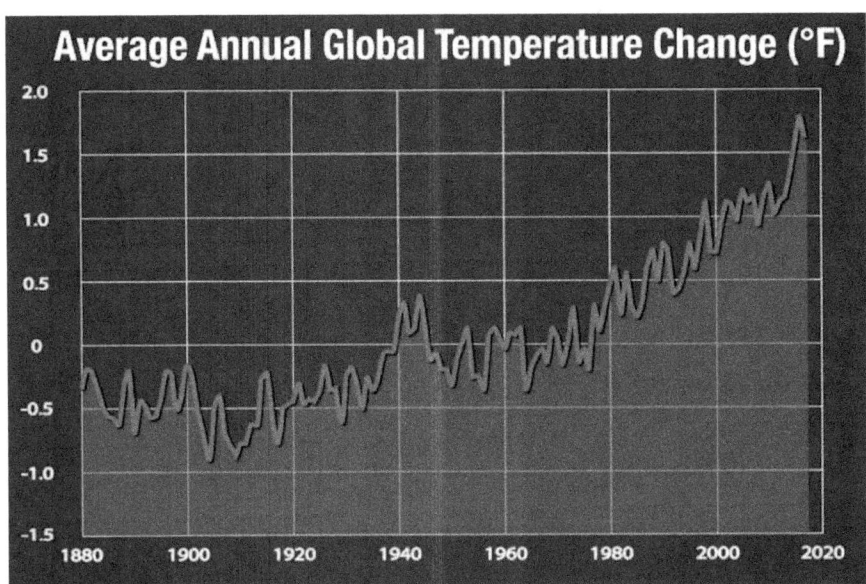

Average Annual Global Temperature Change (°F)

Fig. 2.3. Graph of annual global temperatures, with respect to a baseline from the 19th century (the average of global annual temperatures from 1880-1899). *Credit:* L. Perkins (NASA/GSFC), K. Elkins (USRA), M. Radcliff (KBR Wyle Services, LLC), K. Ramsayer (Telophase), K. Mersmann (KBR Wyle Services, LLC), M. Radcliff (KBR Wyle Services, LLC). https://svs.gsfc.nasa.gov/12828/#section_credits.

Given the global oceans' tremendous size and capacity, it takes massive heat energy to raise even a small amount of Earth's yearly surface temperature. The roughly 2 °F (1 °C) increase in global average surface temperature since the pre-industrial era (1880-1900) might seem small, but it means a significant increase in accumulated heat. The rate of warming since 1981 is more than twice as fast: 0.32 °F (0.18 °C) per decade.

Temperature records worldwide show a clear upward trend, with the last few decades being the warmest on record (Fig. 2.3). The ten warmest years in the historical record have all occurred since 2010. The 2022 surface temperature was 1.55 °F (0.86 °C) warmer than the 20th-century average of 57.0 °F (13.9 °C) and 1.90 °F (1.06 °C) warmer than the pre-industrial period (1880-1900).

2.4 EVIDENCE FROM ICE CORES

Ice cores extracted from polar regions contain historical records of atmospheric composition and temperature dating back hundreds of thousands of years. Snow falling in places like Antarctica or Greenland traps air bubbles, particles, and other substances. As more snow falls, the layers compact into ice, preserving these bubbles and particles. By drilling into the ice and extracting cores, scientists can analyze the air bubbles and other materials trapped in the ice to reconstruct past levels of greenhouse gases, volcanic ash, and other substances. Here's how it works:

The ratios of stable isotopes of oxygen (O) and hydrogen (H) in ice layers can be used to infer past temperatures because these ratios are influenced by the temperature at which the ice formed.

Oxygen comes in several isotopes, the most common being oxygen-16 (^{16}O) and oxygen-18 (^{18}O). When water evaporates, the lighter ^{16}O isotope is more likely to be taken up into the vapor phase, while the heavier ^{18}O isotope is more likely to remain in the liquid phase. As a result, the ratio of ^{18}O to ^{16}O ($^{18}O/^{16}O$) in ice is affected by the temperature of the water from which the ice originated. In colder conditions, oxygen isotopes are more fractionated, resulting in lower $^{18}O/^{16}O$ ratios in the ice. By analyzing the $^{18}O/^{16}O$ ratio in ice cores, scientists can estimate the temperature when the ice was formed.

Similarly, hydrogen has two stable isotopes, protium (^{1}H) and deuterium (^{2}H or D). Deuterium is heavier than protium, so when water vapor condenses into precipitation, the heavier deuterium is more likely to be found in the precipitation. The ratio of deuterium to protium (D/H ratio) in ice is also affected by the temperature at which the ice formed. Lower temperatures result in higher fractionation of hydrogen isotopes, which means lower D/H ratios in the ice. Scientists can also estimate past temperatures by analyzing the D/H ratio in ice cores. These ice cores show a strong correlation between CO_2 levels and temperature, demonstrating that changes in greenhouse gas concentrations influence climate.

In summary, analyzing oxygen and hydrogen isotope ratios in ice cores provides valuable information about past climates. By examining the variations in these ratios in different layers of ice, scientists have developed a detailed picture of Earth's climate history, spanning hundreds of thousands of years. This information is crucial for understanding the Earth's climate system's natural variability and predicting future climate changes.

2.5   ICE MELT AND SEA LEVEL RISE

There are two primary mechanisms by which climate change leads to rising sea levels: the melting of ice sheets and glaciers and the thermal expansion of seawater.

Ice sheets are vast glacial ice expanses covering over 50,000 square kilometers. The two major ice sheets on Earth are located in Antarctica and Greenland. Glaciers are smaller ice masses found on every continent, including Africa and South America. When ice sheets and glaciers melt, they contribute to rising sea levels in two ways. First, the melting ice adds water to the oceans. Second, the melting of ice sheets can affect gravitational forces in a way that raises sea levels. For example, the massive ice sheet of Antarctica exerts a gravitational pull on the surrounding ocean, drawing water towards it. When the ice sheet melts and loses mass, this gravitational pull is reduced, allowing the water to move away and raise sea levels elsewhere.

As the Earth's atmosphere warms due to climate change, the ocean absorbs some heat. When water heats up, its molecules move faster and take up more space, increasing its volume. This process, known as thermal expansion, contributes to rising sea levels.

2.6 CHANGES IN PRECIPITATION PATTERNS

Climate change has wide-ranging effects on rainfall patterns, impacting precipitation frequency, intensity, variability, and distribution. The specific effects of climate change on rainfall can vary by region, but generally, they include:

Changes in Precipitation Intensity: As the atmosphere warms, it can hold more moisture. This can result in heavier and more intense rainfall events, even if the total annual rainfall might not significantly change. This means that when it rains, it can rain harder, leading to flash floods and increased runoff.

Shifts in Rainfall Patterns: Some areas that traditionally receive a lot of rain might experience less, while typically dry areas might see more rain. This can significantly affect local water supplies, agriculture, and ecosystems.

Increased Variability: Places might experience more extended dry periods followed by intense rainfall rather than steady and predictable rain patterns. This can be problematic for agriculture and water management.

Extended Droughts: Some regions, especially those already prone to drought, may experience even longer and more severe droughts due to the effects of climate change. Droughts can have wide-ranging consequences on agriculture, water supplies, and ecosystems.

Changes in Monsoons: The warming atmosphere and changes in land-sea temperature contrasts can influence monsoon patterns, potentially affecting their onset, duration, and intensity. Millions of people in Asia and Africa rely on monsoons for their water needs so any change can have vast implications for agriculture, water resources, and food security.

Polar and Mountain Precipitation: Warmer temperatures can lead to more precipitation falling as rain rather than snow in polar and mountainous regions. This can affect river flows and water resources downstream.

Oceanic Rainfall: Warmer ocean temperatures can lead to more intense tropical cyclones and hurricanes. These storms can produce significant rainfall, leading to flooding in coastal regions.

Melting Glaciers and Ice Caps: While not directly related to rainfall, it is worth noting that melting glaciers and ice caps due to warming temperatures contribute to rising sea levels. Moreover, the freshwater from melting ice can affect ocean salinity and circulation patterns, potentially impacting global precipitation patterns.

Shifts in Atmospheric Circulation: Changes in large-scale atmospheric circulation patterns, like the jet stream, can influence where and when rain falls. For instance, alterations in the El Niño Southern Oscillation (ENSO) due to climate change can affect rainfall patterns globally.

It's essential to understand that these changes are often interrelated and can amplify each other. For instance, a prolonged drought can decrease vegetation cover, leading to reduced local evapotranspiration and even less rainfall, creating a feedback loop.

2.7   EXTREME WEATHER EVENTS

A growing body of evidence links climate change to more frequent and severe extreme weather events, such as hurricanes, heatwaves, droughts, and floods. Numerous scientific studies and assessments conducted over the past several decades use observational data, climate models, and statistical analyses to assess the relationship between climate change and extreme weather events.

Hurricanes and Tropical Cyclones: Research has shown that while the total number of hurricanes may not necessarily increase, the intensity of these storms is likely to increase due to warmer ocean temperatures. Warmer waters provide more energy to fuel storms, making them more powerful and potentially more destructive. Additionally, rising sea levels can exacerbate storm surge impacts from hurricanes.

Heatwaves: The link between climate change and heat waves is one of the most well-established connections. Heat waves become more frequent, intense, and longer-lasting as global temperatures rise. This can significantly impact human health, agriculture, and natural ecosystems.

Droughts: There is also evidence to suggest that climate change is affecting the frequency and severity of droughts. Changes in precipitation patterns, combined with higher temperatures, can lead to more severe and prolonged drought conditions. This can have serious consequences for water supply, agriculture, and ecosystems.

Floods: The relationship between climate change and flooding is complex, as various factors, including precipitation patterns, snowmelt, and land use changes, can influence floods. However, evidence suggests climate change can contribute to more frequent and severe flooding events. For example, extreme rainfall events are expected to become more common, leading to an increased risk of floods.

In addition to these specific examples, numerous other studies and assessments have examined the relationship between climate change and extreme weather events. For example, the Intergovernmental Panel on Climate Change (IPCC) has published several reports that assess the current state of scientific knowledge on this topic. Overall, the evidence suggests climate change contributes to more frequent and severe extreme weather events.

2.8   OCEAN ACIDIFICATION

The absorption of excess CO_2 from the atmosphere by the world's oceans has significant consequences for marine ecosystems and climate systems. This process, known as ocean acidification, occurs when CO_2 dissolves in seawater, resulting in a series of chemical reactions that decrease the ocean's pH levels. Understanding the mechanism behind this phenomenon requires exploring the chemistry involved, the role of the oceans in regulating atmospheric CO_2, and the impacts on marine life and global climate systems.

When the ocean absorbs CO_2, the CO_2 reacts with water (H_2O) to form carbonic acid (H_2CO_3). The H_2CO_3 dissociates into bicarbonate ions (HCO^{3-}) and hydrogen ions (H^+). The increase in hydrogen ions lowers the ocean's pH, making it more acidic. Furthermore, these hydrogen ions can react with carbonate ions (CO_3^{2-}) to form additional bicarbonate ions, reducing carbonate availability. This is crucial because carbonate ions are essential for marine organisms like corals, mollusks, and some plankton species, which use them to build their calcium carbonate ($CaCO_3$) skeletons and shells.

Oceans play a significant role in regulating atmospheric CO_2 levels through a process known as the carbon cycle. The carbon cycle involves the absorption, transformation, and release of carbon in the atmosphere, oceans, and terrestrial ecosystems. The oceans act as a sink for atmospheric CO_2, absorbing approximately 30% of the total CO_2 emissions produced by human activities. This absorption is facilitated by the ocean's natural circulation patterns, which help to distribute CO_2 throughout the water column.

However, the excessive release of CO_2 due to human activities, such as burning fossil fuels, deforestation, and industrial processes, has overwhelmed the ocean's capacity to buffer these changes. As a result, the rate of ocean acidification has increased significantly, with the current rate being about ten times faster than at any point in the last 55 million years.

The impacts of ocean acidification are profound and wide-ranging. As previously noted, the decrease in available carbonate ions due to ocean acidification makes it more difficult for marine organisms to produce their protective structures, leading to weaker shells and skeletons. Coral reefs, vital for supporting diverse marine ecosystems, are particularly vulnerable to ocean acidification. Corals require a specific concentration of carbonate ions to produce their calcium carbonate skeletons. As ocean acidity increases, the availability of carbonate ions decreases, making it harder for corals to build their skeletons. This leads to coral bleaching, where the corals expel the symbiotic algae that live within their tissues, resulting in the loss of their vibrant colors and, eventually, their death.

In addition to its direct effects on marine life, ocean acidification can have broader implications for global climate systems. For example, as the ability of the oceans to absorb CO_2 decreases, more CO_2 remains in the atmosphere, contributing to the greenhouse effect and global warming. This, in turn, can affect weather patterns, sea levels, and the overall stability of the Earth's climate system.

2.9 CONSENSUS AMONG SCIENTISTS

The scientific consensus that human activities are the primary driver of recent global warming is overwhelming and rooted in rigorous research and comprehensive data analysis. An extensive body of scientific literature, supported by numerous peer-reviewed studies, unequivocally establishes the connection between human-induced greenhouse gas emissions and the observed increase in global temperatures.

Empirical Evidence and Observational Data

Multiple lines of empirical evidence and observational data demonstrate that the Earth's climate is changing. Global temperatures have risen by approximately 1.2 °C since the late 19th century, with the last few decades experiencing more rapid warming. The increase in global temperatures coincides with the Industrial Revolution and the subsequent rise in greenhouse gas emissions, primarily CO_2, CH_4, and nitrous oxide (N_2O). These gases trap heat in the Earth's atmosphere, leading to a warming effect.

Climate Models and Projections

As noted previously, climate models are essential tools in understanding the impact of human activities on global warming. These models incorporate various factors, including greenhouse gas emissions, solar radiation, and natural climate variability, to simulate past, present, and future climate conditions. The observed warming trend is accurately reproduced when human-induced greenhouse gas emissions are included in the models. In contrast, models that exclude human activities fail to account for the recent increase in global temperatures.

Peer-Reviewed Scientific Studies

Many peer-reviewed scientific studies support the consensus among climate scientists. One of the most comprehensive analyses of scientific consensus on climate change was conducted by Cook et al. (2013), which reviewed over 12,000 abstracts of peer-reviewed scientific papers published between 1991 and 2011. The study found that 97% of the articles that expressed a position on the cause of global warming endorsed the view that human activities are a significant contributing factor. This overwhelming consensus has only grown stronger in subsequent years as more research has been conducted and our understanding of the climate system has improved.

International Scientific Assessments

International scientific assessments, such as those conducted by the Intergovernmental Panel on Climate Change (IPCC), further validate the consensus on human-induced global warming. The IPCC's Fifth Assessment Report (AR5) concluded that it is "extremely likely" (meaning 95-100% probability) that more than half of the observed increase in global average surface temperature since the mid-20th century is due to human influence. This assessment is based on comprehensive scientific literature reviews and represents the consensus view of thousands of climate scientists worldwide.

The overwhelming majority of climate scientists agree that human activities, mainly burning fossil fuels and deforestation, are the primary drivers of recent global warming. This consensus is grounded in robust empirical evidence, observational data, climate models, peer-reviewed scientific studies, and international scientific assessments. The strength of this scientific agreement underscores the urgent need to take collective and meaningful action to mitigate greenhouse gas emissions and address the impacts of climate change. Failure to do so will have severe and lasting consequences for our planet and future generations.

2.10   CLIMATE MODELS

Although climate models have been briefly discussed earlier, this author is compelled to give the subject a more detailed treatment because this is one of the controversial areas of climate science.

Complex computer models simulate past and future climate scenarios based on observed data and known physical principles. These models play a crucial role in understanding how the Earth's climate system works and how it might change in response to various factors. The following is a brief overview of how these models work:

Physical Principles: Climate models are built upon the fundamental physical principles that govern the Earth's climate system. These principles include atmospheric physics, oceanography, thermodynamics, and more. Scientists use mathematical equations that represent these principles to describe how various components of the climate system interact with each other.

Grid System: To simulate the Earth's climate, the planet's surface is divided into a three-dimensional grid. Each grid cell represents a small portion of the Earth's surface, and the models calculate climate variables (e.g., temperature, precipitation, wind patterns) for each grid cell.

Initial Conditions: Climate models are initialized with observed data, such as current temperature, atmospheric composition, and sea surface temperatures. These initial conditions provide a starting point for the model simulations.

Forcing Factors: Climate models also consider external factors influencing the climate, such as greenhouse gas emissions, volcanic eruptions, solar radiation, and land-use changes. These factors are used as inputs to the models to simulate their effects on the climate.

Time Steps: The models advance in time through discrete time steps, simulating the evolution of the climate system over decades, centuries, or even millennia. The smaller the time step, the more detailed the simulation and the more computational resources required.

Validation and Calibration: Climate models are validated and calibrated by comparing their output to historical climate data. Scientists adjust model parameters to ensure the simulations accurately reproduce past climate observations.

Scenario Runs: Once validated, models can be used to project future climate scenarios. Scientists input different scenarios of future greenhouse gas emissions, land use changes, and other factors to explore possible climate futures.

Ensemble Modeling: Researchers often run ensembles of simulations to account for uncertainties in the models and future scenarios. This involves running multiple simulations with slight variations in input parameters to generate a range of possible outcomes.

These complex computer models are essential tools for understanding how the Earth's climate system functions and predicting future climate changes. They are used by climate scientists, policymakers, and researchers to inform decisions related to climate mitigation and adaptation strategies.

When considering human-induced greenhouse gas emissions, these models consistently predict global warming trends that align with observed temperature changes.

2.11 IMPACT ON ECOSYSTEMS AND BIODIVERSITY

Climate change affects ecosystems and biodiversity in numerous ways, with impacts that can be complex and interconnected.

Temperature Changes: As global temperatures rise, some species may no longer be able to survive in their current habitats. This can lead to shifts in species distributions as they move to cooler areas. Some species may adapt, but others, especially those in specialized or isolated habitats, may face extinction.

Altered Precipitation Patterns: Changes in rainfall patterns can affect water availability, impacting both aquatic and terrestrial ecosystems. Some areas may experience more frequent droughts, while others may experience more intense rainfall and flooding. This can affect plant growth, water quality, and species distribution.

Ocean Acidification: As previously noted, the absorption of excess carbon dioxide by the oceans is causing them to become more acidic. This affects marine life, particularly organisms that rely on calcium carbonate to build shells and skeletons, such as corals, mollusks, and some types of plankton.

Rising Sea Levels: As ice melts and sea levels rise, coastal habitats may be lost, and saltwater may intrude into freshwater ecosystems. This can lead to habitat loss for many species and changes in the composition of plant and animal communities.

Changes in Seasonal Cycles: Changes in temperature and precipitation can shift the timing of natural events, such as plant flowering, animal migrations, and insect emergence. This can disrupt the synchronization between species that depend on each other, for example, between plants and their pollinators.

Extreme Weather Events: More frequent and severe extreme weather events, such as hurricanes, heatwaves, and wildfires, can devastate ecosystems. These events can cause direct harm to species, destroy habitats, and lead to longer-term changes in ecosystem structure and function.

Loss of Ice Habitats: Melting ice in polar regions poses a significant threat to species that depend on ice-covered regions, including polar bears, seals, and penguins.

The impact of climate change on ecosystems and biodiversity is a primary concern for conservationists, scientists, and policymakers.

2.12   FOOD AND WATER SECURITY

Climate change has a profound effect on food and water security. As the global climate system is disrupted, there are direct and indirect repercussions on agricultural productivity, water availability, and the reliability of food systems. Here are some of the key impacts:

Changes in Precipitation Patterns: Reduced rainfall can decrease soil moisture, affecting crop yields. This can also reduce the amount of water available for irrigation. On the other hand, intense rainfall can lead to floods, damaging crops, eroding soil, and disrupting planting and harvesting.

Temperature Increases: Higher temperatures can cause plant heat stress, affecting crop yields and quality. Some crops are more vulnerable than others. Higher temperatures can also increase evaporation from soil and reservoirs, reducing the water available for crops.

Shifts in Growing Seasons: Some regions may experience a shortening or lengthening of their growing seasons, which can affect crop yields. Farmers may need to adapt by changing planting dates or crop varieties.

Changes in Pest and Disease Patterns: Warm temperatures can expand the range of many pests and diseases, exposing crops to threats previously not exposed to.

Loss of Agricultural Land: Rising sea levels can lead to the loss of valuable agricultural land due to salinization. Saltwater intrusion affects soil and freshwater sources, making them unsuitable for many crops.

Water Scarcity: Reduced freshwater availability can impact agriculture and human populations directly. Many regions rely on melting glaciers and snow for freshwater. As these sources diminish, water scarcity becomes a pressing concern.

Impacts on Fisheries: Ocean acidification, caused by increased levels of CO_2, can affect marine life, especially shellfish. Warmer water temperatures can also disrupt marine ecosystems and fisheries.

Supply Chain Disruptions: Extreme weather events can disrupt the transportation and storage of food, leading to increased food prices and reduced access.

Economic Impacts: Reduced agricultural productivity can lead to financial hardships, especially in countries where a large portion of the population relies on agriculture for their livelihood.

Social and Political Impacts: Food and water scarcity can lead to conflicts, forced migrations, and other social disruptions. This can strain national and regional relations and could lead to geopolitical conflicts over resources.

Mitigation and adaptation strategies are essential to address these challenges. This includes developing and adopting drought-resistant crop varieties, improved water storage and irrigation systems, conservation practices, and comprehensive local, national, and global policies.

2.13   HUMAN HEALTH

Climate change can affect human health in numerous ways, including heat-related illnesses, diseases spread by mosquitoes and other vectors, and respiratory problems due to poor air quality.

Heat-related illnesses and deaths: Rising temperatures can lead to more heat waves, which can cause heat stroke, dehydration, and other heat-related diseases. Extreme heat can be particularly dangerous for vulnerable populations such as the elderly, children, and those with pre-existing health conditions.

Air quality and respiratory problems: Climate change can affect air quality by increasing the frequency and intensity of wildfires, which produce smoke and pollutants that can irritate the lungs and exacerbate respiratory conditions like asthma. Additionally, warmer temperatures can increase ground-level ozone, a smog component that can harm respiratory health.

Vector-borne diseases: Climate change can affect the distribution and behavior of disease vectors like mosquitoes and ticks, potentially spreading diseases such as malaria, dengue fever, Lyme disease, and West Nile virus to new areas where they were not previously found.

Water-related illnesses: Extreme weather events, such as hurricanes and floods, can contaminate drinking water supplies and lead to outbreaks of waterborne diseases like cholera and diarrheal infections.

Food safety and nutrition: Climate change can affect food safety by changing the patterns of foodborne pathogens and contaminants. Changes in temperature and precipitation can also affect crop yields and the availability of certain foods, potentially leading to malnutrition and food insecurity.

Mental health impacts: The psychological impacts of climate change, such as anxiety and depression, can be significant, particularly for those directly affected by extreme weather events or who live in vulnerable areas.

Displacement and conflict: Climate change can contribute to displacement and conflict by exacerbating resource shortages and contributing to competition over scarce resources, which can have direct and indirect effects on health. Overall, the impacts of climate change on human health are complex and interconnected, with some populations being more vulnerable than others. Addressing these health impacts requires a multifaceted approach, including reducing greenhouse gas emissions, improving public health infrastructure, and addressing social and economic inequalities.

2.14 SOCIETAL AND ECONOMIC IMPACTS

Climate change has profound impacts on both society and the economy. These impacts are interconnected, as societal changes often lead to economic changes and vice versa. The following are societal impacts.

Migration and Displacement: Rising sea levels and extreme weather events can lead to the displacement of people from their homes. This can result in "climate refugees" who may migrate to other regions or countries, potentially causing social and political tension.

Food and Water Security: Climate change can affect agricultural production, leading to food shortages. This can cause malnutrition and hunger, especially in vulnerable populations. Similarly, changes in precipitation patterns can affect water availability, leading to water scarcity.

Cultural Loss: Some communities may lose their traditional ways of life due to environmental changes. For example, indigenous communities in the Arctic face challenges to their traditional hunting and fishing practices due to melting ice.

The economic impacts include:

Agriculture: Climate change can affect crop yields and livestock production, which can have significant economic consequences, especially in regions where agriculture is a major part of the economy.

Infrastructure Damage: Extreme weather events can damage infrastructure, such as roads, bridges, and buildings, leading to high repair costs. This can be particularly challenging for developing countries that may not have the resources to rebuild.

Energy Production: Climate changes can affect energy production and consumption. For example, rising temperatures can increase the demand for air conditioning, while changes in precipitation can affect hydropower production.

Tourism: Some regions may see a decline in tourism due to climate change. For example, ski resorts may suffer from a lack of snow, while coastal areas may be affected by rising sea levels and hurricanes.

Insurance Costs: As the frequency and severity of extreme weather events increase, insurance costs are likely to rise. This can have implications for homeowners, businesses, and the insurance industry.

2.15 MIGRATION AND CONFLICT

Climate change affects migration in numerous ways, both directly and indirectly. The impacts of climate change on human movement are complex, multifaceted, and often interrelated with other social, economic, and political factors.

Sea-Level Rise: Low-lying coastal areas and island nations are particularly vulnerable to rising sea levels, which can lead to the loss of land and property. This can force communities and populations to relocate to higher ground or other regions or countries.

Extreme Weather Events: Intense hurricanes, floods, droughts, and wildfires, becoming more frequent and severe due to climate change, can destroy homes and livelihoods. Also, people may be temporarily or permanently displaced, seeking refuge in safer areas.

Water Scarcity and Food Security: Changes in precipitation patterns and increased evaporation due to higher temperatures can result in water scarcity. This can affect agriculture, leading to food shortages and loss of income for farmers, compelling people to migrate.

Heatwaves: Prolonged periods of extreme heat can have severe health impacts and may also affect livelihoods, especially for those working in agriculture or outdoors. This can force people to move to cooler regions.

Desertification and Land Degradation: Climate change can contribute to desertification and land degradation, reducing the availability of arable land. This can prompt rural populations to move to urban areas or migrate internationally.

Conflict and Security: The impacts of climate change can exacerbate existing social and political tensions. Also, competition over scarce resources such as water and land can lead to conflict, potentially leading to displacement and migration.

Health Impacts: Climate change can affect human health, potentially impacting infectious disease patterns, heat-related illnesses, and more. Health concerns can be a push factor for migration, as people seek better living conditions and access to healthcare.

Interactions with Other Factors: Climate-induced migration often interacts with other social, economic, and political factors. For instance, pre-existing vulnerabilities such as poverty, lack of access to resources, and governance issues can influence people's ability to adapt to climate change or their decision to migrate.

Climate change also affects conflict through various direct and indirect pathways that can exacerbate existing social, economic, and environmental vulnerabilities. Here are some ways in which climate change can contribute to conflict.

Resource Scarcity: Climate change can lead to resource scarcity, such as water and food shortages, due to changing rainfall patterns, droughts, and extreme weather events. Resource scarcity can lead to competition between different groups, such as farmers and herders, escalating into conflict.

Livelihood Security: Climate change can threaten livelihoods, especially in communities dependent on agriculture or natural resources. Loss of livelihoods can lead to economic instability and increase the likelihood of conflict as people compete for limited resources and opportunities.

Forced Migration and Displacement: Extreme weather events and rising sea levels can lead to forced migration and displacement of people. Displacement can lead to overcrowding in urban areas or refugee camps, increasing competition for resources and services and potentially leading to conflict.

State Fragility and Governance: Climate change can exacerbate state fragility by overwhelming governments' capacity to provide essential services and maintain order. Weak governance can create a vacuum that can be exploited by non-state actors, such as terrorist groups or criminal organizations, which can contribute to conflict.

Social and Ethnic Tensions: Climate change can exacerbate existing social and ethnic tensions, as resource scarcity can lead to competition between different groups. Climate-related migration can also lead to tension between migrants and host communities.

Geopolitical Tensions: Climate change can create geopolitical tensions over resources like water or fertile land. Rising sea levels can lead to disputes over territory and maritime boundaries.

In summary, the effects of climate change on migration are multifaceted and can lead to different types of movements, including internal displacement, cross-border migration, and even international migration. Addressing the impacts of climate change on migration requires a holistic approach, considering environmental, social, economic, and political dimensions. This includes implementing effective climate change mitigation and adaptation strategies, protecting the rights of migrants, and providing support and resources for those affected. As resources become scarcer and extreme weather events more frequent, there may be increased migration and the potential for resource conflict.

2.16   MITIGATION AND ADAPTATION

Mitigation

There are various worldwide actions taken to mitigate climate change, which can be grouped into international agreements, national policies, technological advancements, and grassroots movements:

International Agreements: The Paris Agreement (2015). This is a legally binding international treaty where countries commit to limit global warming to well below 2, preferably 1.5 °C, compared to pre-industrial levels. Countries set their own nationally determined contributions (NDCs) to achieve this goal.

The Kyoto Protocol (1997): The Protocol is an international treaty that commits its parties to reduce greenhouse gas emissions based on the scientific consensus that global warming is occurring, and it is extremely likely that human-made CO_2 emissions have predominantly caused it.

National Policies: Many countries have set their climate action plans and targets, often aligning with their commitments under the Paris Agreement. Some countries have implemented carbon pricing to reduce emissions, like carbon taxing or cap-and-trade systems. Other countries invest heavily in renewable energy sources such as solar and wind and phasing out coal-fired power plants.

Technological Advancements: This includes the development of renewable energy technologies, such as solar panels, wind turbines, and hydroelectric power. Energy storage advancements include battery technologies to support integrating renewable energy into the grid. Research and development of carbon capture and storage technologies to reduce emissions from fossil fuel use is another technological advancement.

Grassroots Movements and Individual Actions: Movements like Fridays for Future and Extinction Rebellion have mobilized millions of people worldwide to demand climate action. Individuals are also taking action by reducing their carbon footprint, such as driving less, using energy-efficient appliances, and reducing waste.

Research and Education: Many academic and research institutions are researching climate change and its impacts. Education campaigns are being conducted worldwide to increase awareness and understanding of climate change and the actions people can take to mitigate it.

These are just some of the worldwide actions being taken to mitigate climate change. It is a complex and multifaceted issue requiring cooperation from governments, industries, and individuals.

Adaptation

Worldwide, numerous climate change adaptation actions are being implemented across different regions, sectors, and communities to reduce vulnerability and increase resilience to the impacts of climate change. Here are some examples:

Adaptation actions in coastal areas include constructing sea walls, storm surge barriers, and other infrastructure to protect against sea level rise and storm surges. Mangroves are planted and restored as natural barriers against storm surges and coastal erosion.

In urban areas, adaptation actions include installing green roofs, permeable pavements, and other green infrastructure to mitigate urban heat island effects and manage floodwaters. Climate-resilient infrastructure includes upgrading roads, bridges, and buildings to withstand extreme weather events.

In agriculture, adaptation measures include developing and cultivating crops that are more resilient to changing climate conditions, such as drought or flood-resistant varieties, and implementing efficient irrigation systems and rainwater harvesting to cope with water scarcity.

Implementing controlled burns, creating firebreaks to prevent wildfires, and helping species migrate to more suitable habitats as climate zones shift are a few of the adaptation strategies in the forestry sector.

In the health sector, enhancing surveillance and monitoring of diseases that may spread due to climate change, such as malaria and dengue, and strengthening healthcare infrastructure and preparedness for climate-related emergencies are examples of adaptation measures.

Global and regional cooperation involves providing financial and technical support to developing countries for climate change adaptation through mechanisms like the Green Climate Fund. It also involves sharing best practices, data, and research among countries and regions to support informed decision-making.

These adaptation actions are part of national and international efforts to tackle climate change, including commitments under the Paris Agreement and the United Nations Framework Convention on Climate Change (UNFCCC).

3
GRASSROOTS ACTIVISM

Grassroots organizations and individuals are crucial in combatting climate change in the US and worldwide. These dedicated individuals and groups work tirelessly to raise awareness, advocate for policy change, and implement practical solutions at the community level. Here's a description of some of the grassroots efforts in the US and globally.

3.1   GRASSROOTS ORGANIZATIONS IN THE US

The following are some of the grassroots organizations in the US.

350.org: Founded by Bill McKibben, 350.org is a global grassroots movement dedicated to addressing climate change. They organize campaigns, demonstrations, and educational events to pressure governments and corporations to reduce carbon emissions.

Citizens' Climate Lobby (CCL): CCL is a nonprofit organization that empowers individuals to advocate for climate solutions by lobbying their members of Congress. They promote a carbon fee and dividend policy to reduce emissions.

Sunrise Movement: Sunrise is a youth-led organization pushing for urgent climate action. They are known for advocating for the Green New Deal, a comprehensive policy framework to address climate change and economic inequality.

Sierra Club: One of the oldest and most prominent environmental organizations in the US, the Sierra Club works at the grassroots level on various environmental issues, including climate change.

The Climate Reality Project: Founded by former Vice President Al Gore, this organization trains individuals worldwide to become climate leaders and advocates for climate action.

3.2   GLOBAL GRASSROOTS EFFORTS

Key examples include:

Fridays for Future: This international movement, inspired by Swedish activist Greta Thunberg, involves students and young people striking from school to demand more decisive climate action. It has sparked protests in countries around the world.

Extinction Rebellion: A global environmental movement, Extinction Rebellion uses non-violent civil disobedience to demand that governments take immediate action on climate change.

Greenpeace: While it's a global organization, Greenpeace's grassroots campaigns are particularly impactful in local communities worldwide. They often engage in direct action and advocacy to address climate issues.

Global Climate Strikes: Coordinated by various organizations, including Fridays for Future and Greenpeace, international climate strikes mobilize millions worldwide in cities to demand climate action from governments and corporations.

Indigenous and Local Communities: Indigenous communities worldwide have been at the forefront of climate action, advocating for protecting their lands and traditional knowledge. Local communities in vulnerable areas often develop innovative climate adaptation strategies.

3.3   STRATEGIES AND TACTICS TO RAISE AWARENESS

Climate activists employ various strategies and tactics to raise awareness about climate change, advocate for policy changes, and push for a transition to more sustainable practices. Some of the key strategies and tactics used by climate activists include:

Lobbying

Lobbying is the practice of advocating for a particular cause, interest, or policy by individuals or groups to influence government decisions. It is an essential part of a democratic society, allowing citizens and organizations to voice their concerns and shape public policies. When it comes to climate change, lobbying can be a powerful tool for bringing attention to the issue, driving policy changes, and encouraging collective action.

Raising Awareness: One of the primary objectives of lobbying is to raise awareness about climate change. Lobbyists work to educate policymakers, the public, and businesses about the science behind climate change, its consequences, and the urgency of taking action. They often utilize various communication channels, including meetings with lawmakers, public events, social media campaigns, and educational materials, to disseminate information and build a broader understanding of the issue.

Influencing Policy: Lobbying also plays a vital role in influencing climate-related policies. Climate change is a complex issue with far-reaching implications, and addressing it requires comprehensive and well-informed policies. Lobbyists work to shape legislation that promotes renewable energy, reduces greenhouse gas emissions, and encourages sustainable practices. They advocate for measures such as carbon pricing, renewable energy incentives, and stricter environmental regulations, all of which can contribute to mitigating climate change.

Promoting Corporate Responsibility: Lobbying efforts extend to the corporate sector, where companies are increasingly being held accountable for their environmental impact. Climate-conscious lobbyists work to persuade businesses to adopt sustainable practices, reduce their carbon footprint, and invest in green technologies. By pressuring corporations to prioritize climate action, lobbyists can accelerate the transition to a low-carbon economy.

Mobilizing Public Support: Lobbying is not limited to influencing policymakers and corporations; it also aims to mobilize public support for climate action. By engaging with communities, organizing rallies, and conducting public outreach campaigns, lobbyists can galvanize grassroots movements that demand change. A well-informed and passionate public can exert significant pressure on governments and businesses to take meaningful steps toward addressing climate change.

Challenges and Controversies: Despite its potential for positive change, lobbying for climate awareness faces its share of challenges and controversies. Some critics argue that powerful interest groups, particularly those tied to the fossil fuel industry, can undermine climate lobbying efforts by funding campaigns and politicians who oppose climate action. Additionally, there is a concern that excessive corporate influence on climate policy can lead to greenwashing – the practice of appearing environmentally friendly without taking substantive steps.

Legal Action

In the face of mounting evidence of climate change and its devastating impacts, climate activists have increasingly turned to the legal system as a potent tool to hold governments and corporations accountable. The courts have become vital arenas for activism, not only to seek redress for environmental damage but also to compel action to mitigate future harm and to enforce policy commitments.

The Public Trust Doctrine: A cornerstone of environmental legal action is the Public Trust Doctrine, a principle tracing back to Roman law that posits specific natural resources as common to all. Activists have invoked this doctrine to argue that governments, as trustees of the environment, must safeguard public resources, such as air and water, for current and future generations. Cases such as Juliana v. United States exemplify efforts to compel government action, asserting a constitutional right to a climate system capable of sustaining human life.

Environmental Impact Assessments: Often, activists resort to the courts to ensure rigorous environmental impact assessments (EIAs) are conducted for projects likely to exacerbate climate change. By litigating the adequacy of EIAs, activists can delay or halt the construction of high- carbon infrastructure, thereby preventing additional greenhouse gas emissions. This legal tactic not only underscores the importance of robust environmental review but also elevates the standards for corporate compliance.

Failure to Act and Policy Enforcement: One of the most direct legal challenges is suing governments for their inaction or inadequate action on climate change. Legal actions may claim that such inaction violates existing climate policies, international agreements, or even human rights. By legally mandating governments to meet their standards, these cases aim to close the gap between policy and practice.

Corporate Misrepresentation and Liability: Climate activists have increasingly scrutinized corporate conduct, bringing lawsuits against companies for misrepresenting their impact on climate or for failing to disclose climate-related risks to investors. Cases against major oil companies, alleging that they concealed their knowledge of the harmful climate effects of fossil fuels, aim to hold them financially and publicly accountable.

Nuisance and Compensation Claims: Activists have utilized traditional tort law, framing climate change as a public or private nuisance. Such litigation seeks to establish a causal link between corporate activities and environmental harm, thereby seeking compensation for affected communities. This strategy not only seeks justice for past and present damages but also aims to deter future negligent behavior.

Human Rights Litigation: Innovatively, climate litigation is increasingly framed around human rights, arguing that a stable climate is integral to fundamental human rights such as health, water, and even life itself. By doing so, activists have broadened the scope of legal accountability, making it a matter of justice beyond environmental regulation.

Constitutional and Administrative Challenges: Activists have brought forward cases arguing that a government's failure to address climate change contravenes constitutional rights to a healthy environment. These actions have a dual purpose: to spur government action and to set legal precedents that embed environmental rights in constitutional law.

International Law and Transboundary Harm: Climate activists have also looked to international law, particularly in instances where actions in one country have harmful environmental impacts on another. These cases underscore the global nature of environmental law and the need for cross-border legal accountability.

Civil Disobedience

Civil disobedience, the active, professed refusal to obey specific laws, demands, and commands of a government or occupying power, has long been a tactic of social movements seeking to precipitate change. In the realm of environmental activism, particularly with respect to climate change, civil disobedience has emerged as a crucial mechanism for drawing public attention to urgent issues and galvanizing political action. This Section explores the role of civil disobedience in the climate movement, assessing its implications, effectiveness, and ethical dimensions.

The Principle of Civil Disobedience: The philosophical underpinnings of civil disobedience lie in the writings of Henry David Thoreau, who argued for the moral imperative to oppose unjust laws and policies. This principle has been adopted by climate activists who contend that inaction and inadequate policies by governments and corporations on climate issues constitute a grave injustice, not only to the current but also to future generations. Civil disobedience in this context becomes a form of non-violent resistance that disrupts the status quo, demanding an urgent response to the climate crisis.

Drawing Attention to Climate Emergencies: The visibility of civil disobedience actions, such as road blockades, sit-ins at political offices, and the obstruction of operations at fossil fuel facilities, is a pivotal aspect of their utility. Such tactics create scenarios that can capture media attention, thereby elevating the climate dialogue in public discourse. By disrupting everyday life, activists highlight the severity of climate threats, illustrating that immediate and drastic changes are necessary to address them.

Mobilizing Public Opinion and Participation: Civil disobedience has the potential to galvanize public opinion and inspire broader civic participation. When citizens witness individuals willing to risk arrest for a cause, it can act as a catalyst for broader societal engagement with climate issues. This form of activism often serves as an entry point for people who may not have previously engaged with environmental issues, expanding the base of the climate movement.

Influencing Policy and Institutional Responses: Through civil disobedience, activists aim to exert pressure on policymakers and institutions to enact substantive changes in climate policy. By creating a sense of crisis, such actions can force political leaders to negotiate and can lead to more ambitious climate targets and timelines. It can also push corporations to reconsider their environmental impact and sustainability practices as they seek to avoid the negative publicity that comes with being the target of protests.

Ethical Considerations: Civil disobedience in the context of climate change raises complex ethical questions. Activists argue that the gravity of the climate crisis justifies the use of disruptive tactics, as traditional avenues of advocacy have often proved inadequate in securing systemic change. However, these actions also raise concerns about the rule of law and the potential for exacerbating social tensions. Proponents of civil disobedience for climate action contend that the moral imperative to prevent catastrophic global warming outweighs these concerns.

Direct Action

Unlike civil disobedience, which inherently involves the deliberate breaking of laws to protest injustice, direct action can be both legal and illegal, ranging from tree-sitting to blockading roads or occupying oil rigs. In the context of climate change, direct action encompasses a wide range of activities designed to confront the sources of carbon emissions directly or raise awareness about the severity of the climate crisis. These actions often involve a physical presence, such as marches, sit-ins, and the occupation of spaces where environmental decisions are being made. The immediacy of direct action is key to its effectiveness, providing stark visual imagery and a human face to the abstract and often distant notion of climate change.

Galvanizing Public and Media Attention: Direct action catalyzes media coverage of climate issues. The dramatic nature of some direct activities, such as activists chaining themselves to equipment at coal-fired power plants or launching kayaks to block oil tankers, captures the public imagination and can break through the noise of the 24-hour news cycle. This heightened visibility can thrust climate change back into the forefront of public discourse, prompting discussions in living rooms, boardrooms, and legislatures.

Mobilizing Public Sentiment and Participation: Beyond the headlines, direct action has the potential to mobilize public sentiment and inspire wider community involvement. Witnessing firsthand the lengths to which individuals are willing to go to protect the environment can be a powerful motivator for others to take action. By demonstrating commitment and urgency, direct action can serve as a rallying cry that resonates with people's intrinsic values, potentially leading to a ripple effect of civic engagement and advocacy.

Influencing Policy and Corporate Behavior: Through direct action, activists seek not only to shine a spotlight on the climate crisis but also to drive substantive policy changes and shifts in corporate behavior. The urgency conveyed through these tactics can push climate change higher on the political agenda and can hold corporations accountable for their environmental impacts. When successful, direct action can lead to tangible outcomes such as the cancellation of projects harmful to the environment, the implementation of more robust regulatory measures, or the divestment of funds from fossil fuel industries.

Ethical and Strategic Dimensions: Direct action, particularly when it skirts the line of legality, invites an examination of ethical considerations. Activists engaging in these tactics argue that the imminent threat of climate change justifies extraordinary measures. The potential for direct action to create inconvenience for the public and to challenge legal boundaries is weighed against the imperative to prevent ecological collapse. Strategically, the use of direct action must be carefully calibrated to ensure that the message is not lost amid accusations of extremism or illegality.

Boycotts and Consumer Pressure

Among these varied strategies to fight against climate change, the use of boycotts and consumer pressure stands out as a powerful tool to effect change. These tactics capitalize on the fundamental principle of demand-driven markets, where consumer preferences can influence the practices of corporations and, by extension, contribute to the larger environmental agenda.

The Mechanism of Consumer Activism: Boycotts, as a form of consumer activism, involves abstaining from purchasing products or services from companies deemed to be environmentally unfriendly or contributing significantly to climate change. The goal is to inflict economic pressure on these entities, compelling them to adopt more sustainable practices. Consumer pressure, more broadly, can also encompass campaigns urging individuals to reduce consumption, demand greener products, and support sustainable businesses, thus shaping market demands and corporate behavior.

Amplifying Climate Change Awareness: Boycotts and consumer pressure campaigns often begin with raising awareness about the link between consumption and climate change. Through education and advocacy, consumers become aware of the carbon footprint of their purchases, from food and fashion to energy and electronics. Awareness is the first step toward behavioral change, and informed consumers are more likely to support sustainable practices and companies that align with their environmental values.

Influencing Corporate Policies and Practices: The real power of boycotts and consumer pressure lies in their ability to influence corporate policies. As companies are inherently motivated by profit, a significant shift in consumer behavior can lead to a reevaluation of business practices. Companies are increasingly responsive to sustainability as market demand, with many adopting greener policies, committing to renewable energy, and reducing waste in response to consumer advocacy.

Catalyzing Broader Social and Economic Change: When successful, consumer-led movements can result in broader social and economic changes. A widespread boycott, especially when it gains media attention, can ignite public debates about climate change and the responsibility of corporations in contributing to or mitigating the crisis. By compelling companies to act, consumer pressure can also indirectly influence supply chains, promoting a more extensive adoption of sustainable practices across industries.

Challenges and Limitations: Despite their potential, boycotts and consumer pressure campaigns face challenges. Not all consumers have the luxury of choice due to economic constraints, and not all companies are susceptible to market pressure. Additionally, there is a risk of oversimplifying complex issues or unfairly targeting certain companies while ignoring others. Moreover, without widespread participation, the impact of boycotts can be limited.

Ethical Considerations and Strategic Application: Ethical considerations play a role in the strategic application of boycotts and consumer pressure. These tactics should aim for constructive engagement, providing a pathway for companies to improve their practices rather than solely punishing them. Furthermore, it is crucial to ensure that such campaigns are inclusive and consider the social implications of their demands, such as the impact on workers and communities reliant on the targeted industries.

International Negotiations and Conferences

The global stage of international climate negotiations and conferences has increasingly become a forum not just for policymakers but also for activists and civil society representatives. Activist participation in these high-level discussions underscores the democratic imperative that those affected by climate change policies, essentially all global citizens, should have a voice in their formation. This Section explores the multifaceted roles that activists play in international climate negotiations, the impacts of their involvement, and the challenges they face in such diplomatic arenas.

Participation and Representation: Activists and non-governmental organizations (NGOs) participate in international climate conferences, such as the United Nations Framework Convention on Climate Change (UNFCCC) Conference of the Parties (COP), in various capacities. They may attend as observers, contribute to the policymaking process through formal and informal channels, or engage in side events and parallel forums designed to influence the official proceedings. Their presence ensures that a wide array of perspectives, especially those from vulnerable populations and indigenous communities, are heard.

Advocacy and Lobbying: Activists use international conferences as platforms to lobby for stronger environmental protections, more ambitious climate targets, and justice for those most affected by climate change. Through direct engagement with delegates, media campaigns, and public demonstrations, they strive to keep the urgency of the climate crisis at the forefront of discussions. Their advocacy efforts often aim to bridge the gap between the science of climate change and the political will needed to address it.

Monitoring and Accountability: One of the critical functions of activist involvement is to monitor the commitments and actions of governments, ensuring transparency and accountability. Activists often act as watchdogs, calling out discrepancies between nations' promises and their actual policies or practices. By maintaining pressure on governments to honor their commitments, activists help to guard against backsliding and promote progress toward meeting international climate goals.

Knowledge Sharing and Expertise: Activists and NGOs frequently possess on-the-ground experience and technical knowledge that can enrich the climate dialogue. They contribute valuable insights into the real-world implications of policy decisions and offer innovative solutions to complex problems. By bringing in the latest research, data, and case studies, they enhance the information base from which decisions are made.

Amplifying Marginalized Voices: International climate negotiations have historically been dominated by the interests of industrialized nations, often overlooking the disproportionate impact of climate change on developing countries and marginalized groups. Activists play a vital role in amplifying the voices of those communities, advocating for climate justice, and ensuring that the principles of equity and fairness are integrated into global climate policies.

Challenges and Obstacles: Despite the essential role of activists, they face numerous obstacles in their efforts to influence international climate negotiations. Access to the negotiation process can be limited, and the complex nature of diplomatic discussions often marginalizes non-state actors. Moreover, activists from poorer nations may lack the resources to attend or sustain a presence at international conferences, which are typically held in locations far from the most affected areas.

Grassroots Organizing

Grassroots organizing is the lifeblood of social movements, and the fight against climate change is no exception. It encompasses efforts from local community groups to international networks, all aiming to enact environmental protection and policy reform through collective action from the bottom up.

Defining Grassroots Climate Activism: Grassroots climate activism is characterized by community-led initiatives that address local environmental issues with a clear understanding of their global impact. It's a movement that starts with individuals and local communities that are directly affected by climate change or by the actions contributing to it. These activists work to raise awareness, change behaviors, influence policy, and implement sustainable practices on a local level, which collectively contribute to the global effort to mitigate climate change.

Mobilization and Empowerment: Grassroots organizing empowers ordinary people to take control of the climate conversation and the actions affecting their environments. Through education and community engagement, grassroots movements mobilize citizens by emphasizing the direct impact climate change has on their immediate surroundings. They are adept at rallying people around specific, local environmental issues, such as opposing a polluting factory, promoting renewable energy projects, or enhancing community resilience against climate-related disasters.

Strategies and Tactics: The tactics of grassroots movements are diverse and context-specific. They often include door-to-door campaigns, local workshops, tree-planting drives, community clean-ups, and establishing community gardens or farmers' markets. They also use petitions, marches, and demonstrations to exert pressure on decision-makers. Technology and social media have become invaluable tools for grassroots organizations to disseminate information, coordinate actions, and amplify their message.

Influence on Policy and Public Opinion: Grassroots movements have the unique ability to influence policy at the local level, which can have cascading effects on regional, national, and international policies. By demonstrating the successful implementation of sustainable practices, grassroots organizations can serve as a model for larger-scale initiatives. They also have a profound impact on public opinion, often serving as a conduit for translating complex climate science into relatable impacts and actionable items for the general public.

Building Networks and Coalitions: A critical strength of grassroots organizing is its ability to develop networks and coalitions that transcend geographic and social boundaries. Grassroots groups often collaborate with like-minded organizations, faith-based groups, schools, and businesses to strengthen their capacity for action and advocacy. These coalitions can create a powerful unified front capable of mobilizing resources and exerting more significant influence on political processes and industry practices.

Challenges to Grassroots Efforts: Despite its potential, grassroots organizing faces numerous challenges, including limited financial resources, burnout among activists, and opposition from powerful interests invested in maintaining the status quo. Additionally, achieving coherent action and policy change can be challenging, given the diversity of voices and perspectives within the movement. Grassroots activists must constantly navigate the balance between local issues and the global context of the climate crisis.

Artistic and Cultural Activism

Artistic and cultural activism in the climate action movement transcends traditional boundaries of advocacy, harnessing the universal language of creativity to inspire change and spark dialogue on environmental issues. Art, in its myriad forms, offers a potent platform for expressing the urgency of the climate crisis, making complex data relatable, and galvanizing public sentiment.

The Power of Artistic Expression: Artistic activism makes the invisible visible and the intangible tangible. Artists translate the abstract threats of climate change into sensory experiences, whether through visual arts, music, performance, literature, or digital media. By conveying the emotional gravity of climate issues, art can cut through the noise of statistics and scientific jargon, striking a chord with diverse audiences and motivating them to engage with the crisis on a personal level.

Cultural Resonance and Storytelling: Cultural activism relies on the deep-seated power of storytelling and the resonance of cultural narratives to influence beliefs and behaviors. Traditional and contemporary stories, legends, and cultural practices can be reinterpreted to include themes of environmental stewardship, resilience, and adaptation. These narratives can foster a sense of identity and purpose, positioning climate action as a continuation of cultural heritage and collective moral responsibility.

Community Engagement and Solidarity: Art and culture have the unique capacity to build community and solidarity by bringing people together for a common purpose. Public art installations, community murals, street performances, and cultural festivals can become rallying points for local environmental actions, uniting individuals from disparate backgrounds in the shared cause of climate activism. In this collaborative space, the act of creation becomes an act of resistance and empowerment.

Influencing the Public and Policy Discourse: Artistic and cultural interventions often serve as powerful conversation starters that can shift public discourse and influence policy. Art exhibits, documentaries, and cultural events can reach policymakers and stakeholders, provoking reflection and, ideally, inspiring legislative or corporate action. Furthermore, such creative works can leave lasting impressions that persist beyond the moment, embedding environmental messages into the social consciousness.

Challenges and Creativity: While artistic and cultural activism is a vibrant component of the climate movement, it faces challenges in measuring impact and sustaining momentum. Moreover, the risk of art being co-opted or commercialized can dilute its message. However, the very nature of creative expression is to adapt and evolve, turning obstacles into opportunities for innovation. By continuously reinventing their approaches, artists and cultural activists ensure that their message remains relevant and compelling.

Education and Youth Engagement

The twin pillars of education and youth engagement stand at the forefront of the climate action movement, propelling it with the vigor of new ideas and the urgency of those set to inherit the Earth.

Educational Imperative for Climate Literacy: Education serves as a cornerstone for understanding and action. A comprehensive climate education equips the young with scientific literacy about climate change, its causes, impacts, and solutions. Schools, universities, and informal educational platforms have the responsibility to integrate climate science into curricula, fostering critical thinking and problem-solving skills among students. This foundational knowledge empowers the youth to make informed decisions and take meaningful actions in their personal and professional lives.

Youth as Catalysts for Action: Young people are not just passive recipients of knowledge; they are active agents of change. Equipped with education on climate issues, youths are leading grassroots campaigns, initiating school and community projects, and starting conversations that challenge the status quo. Their fresh perspectives and inherent stake in the future imbue them with a sense of urgency that is both inspiring and mobilizing, driving the climate movement forward.

Global Movements and Advocacy: The youth climate movement has gained momentum through global networks and social media, culminating in movements like Fridays for Future, inspired by Greta Thunberg. These movements have shown a remarkable ability to mobilize millions of young people worldwide, amplifying their collective voice and demands for immediate and ambitious climate action from leaders. Their advocacy underscores the willingness of the youth to take ownership of their future and hold governments and corporations accountable.

Youth in Policy and Decision-Making: Recognizing the importance of intergenerational justice, efforts are being made to include youth voices in policy and decision-making processes. Young activists are increasingly present in international forums such as the United Nations Climate Change Conferences (COP), where they articulate their concerns and proposals. Their participation is critical in ensuring that policies are not only environmentally sound but also equitable and reflective of the needs of younger and future generations.

Challenges and Empowerment: Despite their energy and commitment, young activists often face challenges such as ageism, limited access to platforms, and the psychological toll of eco-anxiety. Education systems must, therefore, go beyond traditional teaching methods and invest in the emotional and mental resilience of students, preparing them to face these challenges head-on. By providing tools for civic engagement and platforms for expression, educators and leaders can empower young people to lead the charge in the climate battle.

These strategies and tactics are often combined to create a multifaceted approach to address climate change and promote sustainability. Different activists and organizations may prioritize one or more of these methods based on their goals and the specific context in which they operate.

Climate activism strategies have evolved over the years and have seen successes and setbacks. These strategies leverage the power of collective action to address the pressing issue of climate change.

3.4 SUCCESSFUL OUTCOMES

Global Awareness

Climate activism has been a driving force in catapulting the issue of climate change from the fringes of public consciousness to the center stage of global discourse. Through a combination of relentless advocacy, innovative campaigning, and strategic mobilization, climate activists have successfully highlighted the immediacy and severity of the environmental crisis facing the planet.

Mobilization of Mass Movements: One of the most visible successes of climate activism has been the ability to mobilize mass movements. Iconic events such as Earth Day, global climate strikes, and the People's Climate March have demonstrated the widespread concern and demand for action across different societies. These mass mobilizations have not only drawn media attention but have also served as a stark visual representation of the collective demand for immediate action on climate issues.

Strategic Use of Media and Communication: Climate activists have adeptly used both traditional and new media to spread their message. The strategic framing of climate change as a human story rather than a distant scientific phenomenon has made the issue more relatable and compelling. Documentaries, viral social media campaigns, and celebrity endorsements have managed to penetrate the daily conversations of millions, making climate change a topic of household discussion.

Elevation of Indigenous and Frontline Voices: A significant contribution of climate activism has been its inclusivity in elevating the voices of indigenous peoples and those on the frontlines of climate impacts. By bringing these testimonies to global platforms, activists have created a powerful narrative around climate justice, emphasizing that those least responsible for climate change often face the most severe consequences. This has fostered a broader understanding of climate change as a multifaceted issue that intertwines with human rights, social equity, and intergenerational justice.

Influence on Political Agendas and International Policy: Activists have been instrumental in pushing climate change up the political agenda. Through sustained advocacy and lobbying, they have pressured governments and international bodies to prioritize climate action. The Paris Agreement of 2015 stands as a testament to the impact of sustained global advocacy, with near- universal agreement on the need to limit global warming. Climate activism has ensured that international dialogues continue to reflect the urgency of the issue, contributing to policies that are increasingly aligned with scientific recommendations.

Education and Youth Empowerment: Climate activism has profoundly shaped educational agendas, leading to the integration of climate education in schools around the world. By empowering young people who are particularly active in the climate movement, activists have ensured a pipeline of informed, passionate individuals ready to continue the fight for the planet. The involvement of youth has not only brought fresh energy to the movement but has also underscored the time-sensitive nature of climate action.

Challenges and Continuing the Momentum: Despite these successes, climate activists face the challenge of overcoming political inertia, corporate interests, and a looming climate fatigue among the public. To maintain and increase global awareness, activism must continually innovate and adapt to the changing landscape of communication and engagement.

Policy Changes

As the impacts of climate change become increasingly severe, activists have played a pivotal role in pushing governments to adopt and implement robust climate policies. The persistent efforts of individuals, communities, and organizations across the globe have spurred political action to address what is arguably the defining challenge of our time.

Grassroots Mobilization and Public Pressure: Activists have employed grassroots mobilization to build public support for strong climate policies. By organizing marches, rallies, and public demonstrations, they have created a visible and vocal demand for action that politicians find hard to ignore. The global climate strikes, spearheaded by youth activists like Greta Thunberg, have been particularly effective in capturing public attention and galvanizing a broad base of support for policy change.

Litigation as a Tool for Accountability: Climate litigation has emerged as a powerful tool in activists' arsenals. By taking governments and corporations to court, activists have sought to hold them accountable for their contributions to climate change or their failures to act on existing policies and commitments. Landmark cases, such as the Urgenda case in the Netherlands, have resulted in court mandates for governments to accelerate emission reduction efforts, demonstrating the potential of legal action to compel policy implementation.

Lobbying and Policy Advocacy: Activists have become increasingly sophisticated in their direct engagement with policymakers. Through lobbying efforts, they have provided scientific evidence, economic arguments, and moral imperatives for adopting comprehensive climate policies. They have also offered practical policy solutions, ranging from carbon pricing to renewable energy incentives, helping to bridge the gap between advocacy and policy formulation.

International Pressure and Diplomacy: On the international stage, activists have used the power of global opinion to pressure governments into action. They have played influential roles at international negotiations, such as the United Nations Framework Convention on Climate Change (UNFCCC) conferences, to push for ambitious global agreements. Activist pressure was instrumental in the lead-up to the adoption of the Paris Agreement, encouraging countries to make more substantial commitments to reduce greenhouse gas emissions.

Coalitions and Alliance-Building: Recognizing the multifaceted nature of climate change, activists have built coalitions that transcend traditional environmental groups, including labor unions, faith communities, business leaders, and health professionals. These broad-based alliances have been effective in pushing for climate policies that address a range of societal needs and benefits, making it clear that climate action can serve multiple public interests.

Amplification through Media and Public Discourse: Activists have skillfully utilized media, both traditional and digital, to shape public discourse on climate change. By framing the climate crisis as an urgent, human-centric issue, they have succeeded in capturing the media's attention and, by extension, the public's interest. This media savvy has kept climate change in the news, increasing the political costs of inaction.

Challenges and the Path Forward: While there have been significant successes, activists face the ongoing challenge of ensuring that policies are not only adopted but also fully implemented and enforced. The complexity of climate change requires policies that are adaptive and resilient to new scientific findings and socioeconomic shifts.

Divestment

The fossil fuel divestment movement is a strategic pillar in the global fight against climate change, targeting the financial underpinnings of the fossil fuel industry. By urging investors to withdraw their capital from companies involved in the extraction and production of fossil fuels, activists aim to devalue and delegitimize the economic engines that significantly contribute to global carbon emissions.

Origins and Ethical Underpinnings: The divestment movement took inspiration from the successful divestment campaigns against South African apartheid in the 1980s. It roots itself in the ethical stance that it is morally indefensible to profit from the climate crisis. This moral argument is bolstered by a practical message: investments in fossil fuels are not only environmentally unsustainable but also increasingly financially risky, given the global shift towards renewable energy sources.

Strategies of the Movement: At its core, the fossil fuel divestment movement has focused on institutions such as universities, religious organizations, pension funds, and governmental bodies, urging them to end their financial support for fossil fuel companies. Through petitions, public demonstrations, and negotiations, activists have put pressure on trustees and decision-makers to align their investments with their stated values and the urgent need to combat climate change.

The movement also leverages public shaming and reputation-focused campaigns to influence the divestment decisions of high-profile investors. The message is clear: if you claim to be concerned about climate change, your investments must reflect that concern. This strategy has proven effective in swaying public opinion and, consequently, the decisions of institutional investors.

Impacts on the Industry and Investment Landscape: The divestment movement has achieved notable successes, with over 1,300 institutions across the globe committing to divest from fossil fuels, amounting to trillions of dollars in assets. This has sent a strong signal to the market, questioning the long-term viability of fossil fuel investments and highlighting the growth potential of renewable energy sectors. As the movement gains traction, it has begun to influence the investment industry, encouraging the growth of "green" funds and portfolios that explicitly exclude fossil fuel companies. This not only redirects capital towards renewable energy and sustainable projects but also reshapes the public perception of fossil fuels as a sunset industry.

Challenges and Criticisms: Despite these achievements, the movement faces criticism. Skeptics argue that divestment simply shifts ownership of fossil fuel stocks without directly impacting corporate behavior or reducing emissions. Moreover, in the short term, the sale of shares may not significantly impact large fossil fuel companies that continue to enjoy demand for their products.

However, advocates of divestment counter that the primary goal of the movement is to revoke the social license of fossil fuel companies and stigmatize their role in the economy while also pushing for the societal and policy change needed to reduce emissions.

Role of Divestment: The divestment movement is part of a larger suite of strategies required to address climate change. While it is not a silver bullet, it plays a crucial role in the transition to a low-carbon economy by undermining the financial stability of fossil fuel companies and encouraging the flow of capital towards sustainable alternatives.

Renewable Energy Growth

The swift expansion of renewable energy sources worldwide is not solely a technological or economic phenomenon; it is also a testament to the power of activism. Grassroots campaigns, public advocacy, and policy lobbying by environmental activists have significantly contributed to creating a conducive atmosphere for renewable energy adoption.

Cultural and Behavioral Shifts: Activists have played a critical role in altering public perception, making the adoption of renewables not just a policy preference but a cultural imperative. Through educational campaigns, activists have raised awareness about the benefits of renewable energy, such as reduced carbon emissions and improved public health outcomes. This awareness has increased consumer demand for clean energy and has motivated individuals to adopt renewable technologies at home, such as solar panels and heat pumps.

Policy Advocacy: Activists have tirelessly lobbied for policies that favor the development and integration of renewable energy. Their efforts have been instrumental in securing subsidies for solar and wind power, mandating renewable energy targets, and implementing carbon pricing mechanisms. Such policies have lowered the barriers to entry for renewable energy providers and have made renewables more competitive with fossil fuels.

Influence on Financial Flows: A significant area where activism has been effective is in the redirection of financial flows away from fossil fuels and towards renewables. Campaigns for fossil fuel divestment and green investment have not only deprived the fossil fuel industry of capital but also made significant funds available for renewable energy projects. Activist pressure has prompted investors and financial institutions to recognize the long-term viability and ethical soundness of renewable energy investments.

Direct Action Against Fossil Fuels: Activism has directly impeded the expansion of fossil fuel infrastructure through protests, legal challenges, and direct actions. By opposing coal mines, fracking sites, and pipelines, activists have created economic and political pressure that makes renewables comparatively more attractive and viable. These actions have had the dual effect of hindering fossil fuel development and signaling market preferences for clean energy sources.

Support for Innovation and R&D: Activists have advocated for increased funding and support for research and development in renewable energy technologies. By pushing for government and private sector investment in innovation, activists have contributed to breakthroughs that have made renewable energy more efficient and cost-effective, thereby accelerating its adoption.

Building Alliances: The renewable energy movement has benefited from alliances between activists, the private sector, academia, and policymakers. Activists have built coalitions that span various segments of society, creating a broad base of support for renewables. These alliances have been crucial in enacting legislation, influencing business practices, and generating a groundswell of support for clean energy.

Challenges and Continuing Advocacy: Despite these advances, the transition to renewable energy is not without its challenges. Issues such as grid reliability, energy storage, and the intermittency of some renewable sources remain. Activists are crucial in pushing for solutions to these challenges, ensuring that the momentum behind renewable energy does not wane.

Corporate Accountability

In recent decades, a seismic shift has occurred in the corporate world, partly due to the unrelenting pressure exerted by activists. These advocates for the environment have been unwavering in their demand for companies to adopt sustainable practices.

Direct Action and Public Campaigns: Activists have used direct action to spotlight corporate practices that harm the environment, organizing boycotts, protests, and social media campaigns to draw public attention. These campaigns often harness the power of storytelling, juxtaposing the damages of corporate activity against the broader narrative of climate change and environmental degradation. The resulting public scrutiny can tarnish brands, impacting their reputation and bottom line, which often compels companies to change their practices.

Shareholder Activism: Another powerful tactic has been shareholder activism. By purchasing shares and gaining a voice in shareholder meetings, activists have pushed resolutions that direct companies toward sustainability. This strategy not only creates internal pressure but also sends a signal to the market that sustainability is a priority for investors, thus influencing corporate priorities and attracting like-minded shareholders who value corporate responsibility.

Engagement and Collaboration: Some activists choose to engage directly with corporations, offering insights into how they can improve their sustainability. This collaborative approach often involves dialogues with company leaders, providing them with sustainability plans and helping them understand the economic benefits of adopting greener practices. This method can be particularly effective as it builds a constructive relationship between activists and businesses, fostering a culture of ongoing improvement.

Supply Chain Transparency: Activists have also emphasized the importance of transparency in corporate supply chains. By exposing environmental abuses in supply chains and demanding accountability, they force corporations to scrutinize and often overhaul their procurement processes. This not only impacts the immediate suppliers but also reverberates along the entire supply chain, leading to broader changes in industry standards.

Legal and Regulatory Advocacy: Legal pressure has been used to compel corporations to adopt more sustainable practices. Activists have initiated lawsuits and lobbied for stricter environmental regulations, which has made unsustainable corporate practices more costly and risky. By shaping the legal landscape in which companies operate, activists have made sustainability a legal compliance issue, not just a discretionary choice.

Divestment and Investment: The movement to divest from companies involved in environmentally harmful activities and to invest in sustainable enterprises has gained considerable momentum. Activists have convinced major investment funds, including university endowments and pension funds, to withdraw their investments from these companies. The message is clear: unsustainable companies are not only damaging the planet but are also increasingly seen as risky financial investments.

Consumer Education and Demand: Finally, activists have worked tirelessly to educate consumers about the environmental impacts of the products they buy. By promoting an informed consumer base that demands sustainable products, they have created a market incentive for companies to adopt green practices. The rise in consumer demand for "eco-friendly" and "sustainable" products testifies to the success of these educational efforts.

3.5 SETBACKS

Political Opposition

Climate activists, in their quest to address the existential threat of climate change, routinely encounter significant political opposition. This opposition can emerge from various sectors, including, but not limited to, governments, political parties, industry groups, and lobbyists who perceive climate action as contrary to their interests.

Nature of Political Opposition: The political opposition to climate activism often originates from deeply entrenched economic interests, particularly those allied with the fossil fuel industry. Political resistance can take various forms, such as the propagation of climate denialism, the implementation of policies that favor fossil fuel consumption, and the dismantling of environmental regulations. Often, such opposition is bolstered by misinformation campaigns that downplay the urgency of climate change or exaggerate the economic costs of transitioning to a low-carbon economy.

Strategies of Climate Activists: Despite the headwinds, climate activists deploy a multifaceted arsenal of techniques to push back against political opposition.

(1) Activists organize mass mobilizations to show public support for climate action. Large-scale demonstrations, marches, and coordinated global events like the Global Climate Strike serve to challenge political

inaction and create a sense of urgency.

(2) Climate litigation has become a tool for activists to hold governments and corporations accountable. By taking legal action, activists can force political actors to adhere to their own environmental laws and commitments under international agreements.

(3) Activists build coalitions across civil society, including with businesses, faith groups, and other social movements, to create a broad front that can exert collective pressure on political systems.

(4) Some activists engage directly with the political process, supporting climate-conscious candidates or running for office themselves to effect change from within political institutions.

(5) Dialogue with political leaders and policymakers is another avenue for activists to advocate for climate action, providing scientific evidence and policy solutions to guide decision-making.

(6) Activists collaborate with scientists and policy experts to develop detailed, feasible policy proposals that address climate change, thereby providing ready-to-implement solutions for political leaders.

(7) By engaging with international bodies and leveraging international agreements, activists can create pressure on national governments to commit to stronger climate action.

Challenges and Persistence: Climate activists often find themselves up against powerful interests with significant resources at their disposal. Industry groups can outspend activist campaigns, and political leaders may be unwilling to take the necessary actions due to fear of electoral repercussions or economic impacts. Additionally, the divisive nature of climate politics can lead to polarization, making consensus and action more difficult.

Despite these challenges, the persistent efforts of activists have led to significant victories. The Paris Agreement, for instance, is a testament to the global demand for concerted action against climate change, influenced heavily by activist pressure. Moreover, activists have succeeded in changing the discourse around climate change, making it a central issue in political debates and elections.

Corporate Resistance

The battle to mitigate climate change is not only scientific and cultural but also deeply political and economic. At the heart of this struggle are major fossil fuel companies whose business models and profit margins are inherently tied to the continued extraction and use of coal, oil, and natural gas. Despite growing environmental awareness and a surge in clean energy technologies, these corporations exert considerable effort to resist stringent climate action and lobby against regulations that threaten their interests.

Methods of Resistance: Fossil fuel companies invest heavily in lobbying efforts to influence political decisions. By employing a legion of lobbyists, these corporations work to shape energy policies in a manner that favors the continuation of fossil fuel dominance, often under the guise of energy independence or economic pragmatism.

Political campaign contributions are a strategic tool used by the fossil fuel industry to gain political favor. By financially supporting candidates sympathetic to their agenda, these companies ensure that their voice is heard in legislative bodies, often resulting in the stalling or watering down of ambitious climate legislation.

The spread of misinformation is a critical part of the fossil fuel industry's strategy. These companies have funded research that casts doubt on climate science and have supported think tanks and public relations campaigns that question the severity of climate change and the efficacy of renewable energy.

By placing industry veterans in key governmental regulatory positions, fossil fuel companies work to ensure that the rules governing their operations are favorable to them. This "revolving door" between industry and regulatory agencies often leads to a softening of environmental oversight.

When regulations that threaten fossil fuel interests are passed, these companies do not hesitate to mount legal challenges. Through lawsuits, they seek to delay, block, or overturn regulations designed to reduce carbon emissions and promote clean energy.

The industry frequently cites economic concerns, arguing that strict climate regulations will lead to job losses, higher energy prices, and economic instability. These arguments are often amplified by their allies in political circles and by certain media outlets.

Implications for Climate Policy: The resistance of fossil fuel companies poses significant challenges for climate policy. Efforts to transition to a low-carbon economy are often slowed down or derailed. The domestic politics of countries with powerful fossil fuel lobbies can hamper meaningful progress in international climate negotiations.

This resistance also influences the public discourse around climate change, creating an environment where myths and uncertainties are perpetuated. It can lead to a divided public opinion, where the urgency of climate action is recognized by many but acted upon by few due to the politicization of the issue.

Furthermore, the delay in action enforced by these companies' resistance could lead to more severe consequences of climate change. As scientific projections grow increasingly dire, the cost of inaction — often downplayed by the fossil fuel industry — becomes ever more apparent in the form of extreme weather events, economic losses, and social disruption.

Complexity of the Issue

Climate change is a complex phenomenon, encompassing a vast array of scientific, economic, and social dimensions. Activists dedicated to mitigating climate change are often faced with the daunting task of conveying this complexity to the public and policymakers in a way that is both understandable and compelling. This challenge is critical to garner support for the necessary actions to address the crisis.

The Complexity of Climate Science: At the heart of the challenge is the science of climate change itself, which involves intricate systems and feedback loops that are not easily distilled into soundbites. Scientific terminology and data can be impenetrable to the layperson, creating a barrier to understanding and engagement. Activists must, therefore, find a balance between scientific accuracy and accessibility in their communication strategies.

Psychological Barriers to Communication: The abstract and often intangible nature of climate change projections can make it difficult for individuals to perceive the immediate risks and to motivate behavior change. Activists are tasked with overcoming cognitive biases such as the tendency to prioritize short-term rewards over long-term consequences and the difficulty in grasping the global scale and collective impact of individual actions.

The Challenge of Scale and Relevance: Communicating the global scale of climate change while making it relevant to individual communities is another hurdle. Activists strive to link the global phenomenon to local impacts and individual experiences. This localization of climate issues helps people understand how climate change affects them directly and what they can do about it.

Overcoming Misinformation: Activists also contend with a deluge of misinformation that clouds public understanding of climate change. Industry-funded denial campaigns and politicized rhetoric have contributed to a polarized environment where facts are often contested. Cutting through this noise requires persistent clarification, fact-checking, and the strategic use of reputable sources.

Narrative and Emotional Engagement: To effectively communicate the complexity of climate change, activists often turn to storytelling and emotional engagement. By sharing personal stories of those affected by climate change or using powerful imagery and narratives, activists can create an emotional connection that transcends data and statistics.

The Spectrum of Solutions: Another dimension of complexity lies in the range of potential solutions to climate change. From individual lifestyle changes to systemic policy reforms, the spectrum is broad and multifaceted. Activists face the task of promoting a coherent set of actions that individuals and governments can undertake without overwhelming or confusing their audience.

Strategies for Effective Communication: Activists use metaphors, analogies, and clear visuals to break down complex scientific concepts into understandable and memorable pieces of information.

By sharing relatable stories and case studies, activists can convey the urgency of climate change in a way that resonates on a personal level.

Making the global issue of climate change relevant to local contexts helps individuals understand the practical implications for their own lives and communities.

Activists facilitate discussions that allow for questions, address doubts, and involve the audience in a learning process about climate change.

To maintain motivation and hope, activists highlight incremental progress and showcase successful examples of climate action.

Interests of Different Stakeholders

Climate activism is not a solitary endeavor but a complex interplay of various actors and their competing interests. Governments, businesses, and individuals all hold stakes in the outcomes of climate action, and their interests can be aligned or conflicting. Activists, in their pursuit of environmental sustainability, must navigate this intricate terrain, striving to harmonize these interests to foster collective action against climate change. The challenge begins with a clear understanding of the stakeholders involved.

(1) Governments: Interested in economic growth, energy security, and maintaining public welfare while also being accountable to a populace that increasingly demands sustainable practices.

(2) Businesses: Typically prioritize profitability and market stability but face mounting pressure to adopt environmentally friendly operations and contribute to sustainable development goals.

(3) Individuals: Have varying levels of environmental consciousness, balancing their concern for the planet with immediate economic and social needs.

Strategies for Balancing Interests: Activists employ a range of strategies to bridge the gap between conflicting interests and work towards a common goal:

(1) Advocacy and Awareness Campaigns: By raising awareness about the impacts of climate change, activists strive to create a shared sense of urgency that can transcend individual stakeholder agendas.

(2) Policy Influence: Activists engage in policy advocacy to encourage governments to create regulations that both protect the environment and offer incentives for businesses and individuals to follow suit.

(3) Collaboration and Dialogue: Activists often facilitate dialogue between stakeholders to find common ground and mutually beneficial solutions. This involves understanding the constraints and opportunities each stakeholder faces.

(4) Market Mechanisms: Leveraging market-based solutions, such as carbon trading or green investment funds, activists can align business interests with climate goals, showing that profitability and sustainability can go hand-in-hand.

(5) Legal Action: In some cases, activists resort to litigation to ensure environmental regulations are enforced, protecting the collective interest over individual corporate agendas.

(6) Grassroots Movements: Empowering local communities and individuals to take action provides a bottom-up approach to climate activism, building a broad base of support for environmental initiatives.

Challenges and Adaptation: Balancing these interests is fraught with challenges. There can be resistance to change due to entrenched economic models, political inertia, or individual habits. Moreover, activists may face pushback from powerful lobbies that influence policy against environmental interests.

To adapt, activists must be flexible and innovative in their approaches, employing a mix of persuasive tactics, incentives, and educational tools. They must also stay informed of scientific developments, economic trends, and political shifts to remain effective.

The Role of Compromise: Compromise is often necessary in the world of climate activism. While activists aim for the ideal, they must also acknowledge the practicalities of what is achievable in the short term. This may involve supporting transitional solutions, such as the gradual phase-out of fossil fuels, or accepting incremental policy changes that pave the way for more significant future action.

3.6   POWER OF COLLECTIVE ACTION

In the face of climate change, a threat that disregards borders and scoffs at individual efforts, the necessity for collective action has never been more pronounced. The climate change movement has underscored the power of unity, demonstrating that when individuals come together, their combined strength can catalyze significant environmental transformations.

Collective action occurs when individuals with a common interest band together to achieve a goal. In the context of climate change, this means rallying diverse groups around the shared objective of mitigating the crisis and fostering a sustainable future. This unity not only multiplies individual influences but also creates a ripple effect that can sway the tides of public opinion and political will.

Amplification of Voices: One of the fundamental impacts of collective action is the amplification of voices. When people speak as one, their message resonates louder and clearer. The climate movement has seen a surge in collective platforms, from international organizations to grassroots campaigns, where a chorus of concerned citizens and activists can express their urgency for change.

Policy Influence and Political Mobilization: Collective action is a potent tool for influencing policy. Mass movements can lead to political mobilization, driving legislators to take notice and act. The climate movement has successfully organized marches, petitions, and campaigns that have pressured governments worldwide to prioritize climate policies, commit to emission reduction targets, and fund renewable energy initiatives.

Global Solidarity and Movement Building: The climate crisis knows no boundaries, and neither does the response it has galvanized. Collective action has fostered a sense of global solidarity, with movements like Fridays for Future and the People's Climate March showcasing international unity in climate activism. This global aspect of the movement transcends national policies, pushing for a cooperative approach to addressing the crisis.

The Synergy of Diverse Stakeholders: The convergence of diverse stakeholders characterizes collective action in the climate movement. Scientists, youth, indigenous communities, and businesses join forces, bringing a diversity of perspectives and solutions. This diversity enriches the movement, allowing for holistic and innovative approaches to combating climate change.

Cultural Shift and Behavioral Change: Beyond policy and innovation, collective action has the transformative power to shift cultures and change behaviors. The solidarity in the movement creates new norms and values centered around sustainability. It has the power to influence consumer behavior, encourage sustainable lifestyles, and create a culture that holds environmental stewardship as a core value.

The Role of Technology and Social Media: Modern collective action is significantly bolstered by technology and social media. These tools have democratized activism, allowing for instant communication, coordination, and mobilization of individuals across the globe. Campaigns can go viral, movements can be organized with lightning speed, and messages can reach global audiences instantaneously.

Challenges and Resilience: Despite its strengths, collective action faces challenges such as maintaining momentum, ensuring inclusivity, and overcoming disinformation. The climate movement has, however, shown resilience, adapting to these challenges with creativity and resolve. This adaptability is crucial to sustaining the impact of collective efforts.

4

POLITICAL AND SOCIAL BARRIERS

Climate activists face many political and social barriers that can make their efforts challenging and sometimes even discouraging. These barriers stem from multiple factors, including economic interests, political ideologies, international relations, institutional barriers, social and cultural barriers, opposition from friends and family, legal and security challenges, lack of resources, and activist fatigue.

4.1 ECONOMIC INTERESTS

Economic interest is one of the most significant political barriers to climate activists as it plays a crucial role in shaping policies and influencing decision-makers. Many industries, such as fossil fuels, agriculture, and transportation, have a vested interest in maintaining the status quo and resisting regulations that could limit their emissions and impact their profits. This economic interest is often at odds with the goals of climate activists, who are pushing for urgent and comprehensive action to mitigate climate change.

One of the most evident examples of economic interest as a barrier to climate action is the fossil fuel industry. This industry is one of the largest contributors to greenhouse gas emissions, and it has a long history of lobbying against regulations and funding climate change denial campaigns. Many politicians receive funding from fossil fuel companies, which can influence their policy decisions. The sector also employs many people and contributes to the economy, making it a powerful political force. In many cases, the interests of the fossil fuel industry are directly opposed to the goals of climate activists, and this conflict is a major barrier to achieving meaningful climate action.

Another example of economic interest as a barrier to climate action is the agriculture industry. The agriculture industry is a significant contributor to greenhouse gas emissions, and it has resisted efforts to regulate its emissions and promote more sustainable farming practices. The industry has significant political power and uses this power to influence policy and protect its interests. This resistance to change is a substantial obstacle for climate activists who are calling for a shift towards more sustainable and climate-friendly agriculture.

The transportation industry is also a significant contributor to greenhouse gas emissions, and it has resisted efforts to regulate emissions and promote alternative forms of transportation. The industry has significant economic and political power, and it uses this power to protect its interests and resist change. This resistance to change is another major obstacle for climate activists who are calling for a shift towards more sustainable and climate-friendly transportation options.

4.2   POLITICAL IDEOLOGIES

Political ideology affects how individuals perceive and prioritize climate change. Generally, those on the left (liberals and progressives) tend to view climate change as a severe and urgent problem that requires immediate action. They advocate for stronger environmental regulations, renewable energy investments, and international cooperation to reduce greenhouse gas emissions. Those on the right (conservatives and libertarians) are more likely to be skeptical of the science behind climate change. They oppose measures supported by liberals and progressives, arguing that they would harm the economy and individual freedoms.

This difference in public opinion on climate change is instrumental in shaping policy responses to climate change all around the world. In the United States, for example, climate change has become a highly partisan issue, with Democrats generally supporting more decisive climate action and Republicans often opposing it. This division is reflected in the positions of elected officials, media coverage, and public opinion. For example, in the 2020 presidential election, climate change was a major point of contention between Democratic candidate Joe Biden and Republican incumbent Donald Trump.

In Australian politics, climate change has also been a contentious issue, with divisions emerging between the conservative Liberal Party, which has been criticized for its lack of climate action, and the more progressive Labor Party, which supports more robust climate policies. These divisions have played out in the media and have influenced public opinion.

In Brazil, the issue of climate change is often linked to deforestation in the Amazon rainforest. The government of President Jair Bolsonaro has been criticized for its policies that have led to increased deforestation. At the same time, environmentalists and indigenous communities have called for stronger action to protect the rainforest.

While there is generally more consensus on the need for climate action in Europe, there are still divisions between countries and political parties. For example, some Eastern European countries have been reluctant to commit to the European Union's climate goals, citing concerns about the economic impact on their coal-dependent economies.

On the international stage, political ideology can influence a country's willingness to participate in global climate agreements and initiatives. For example, the United States withdrew from the Paris Agreement under President Donald Trump, a conservative, but rejoined under President Joe Biden, a liberal. Similarly, countries with left-leaning governments are generally more likely to take a leading role in international climate negotiations.

4.3   INTERNATIONAL RELATIONS

International relations can pose a significant barrier to effective climate action in several ways. Countries may have conflicting interests that hinder international cooperation on climate issues. For example, a country with substantial oil reserves may be reluctant to agree to initiatives that would reduce fossil fuel consumption, as this would negatively impact its economy.

Inequality in responsibility and impact can also pose a barrier to climate change solutions. Countries that have contributed the most to greenhouse gas emissions are often not the ones that will be most severely impacted by climate change. This can create tension between developed and developing countries, with the latter constantly arguing that they should not have to bear the same responsibility for mitigating climate change as the former.

Solving global problems requires binding and enforceable global solutions. While there have been international agreements aimed at addressing climate change, such as the Paris Agreement, these agreements are often non-binding and lack enforcement mechanisms. This means that countries may not be held accountable if they fail to meet their emissions reduction targets.

Countries are at different stages of development and may have different capacities for implementing climate change mitigation measures. Developing countries may lack the resources and technology needed to transition to cleaner energy sources and may prioritize economic development over environmental protection.

Geopolitical tensions can also impede international cooperation on climate change. For example, conflicts between countries can distract from efforts to address climate change and make it more challenging to reach a consensus on international agreements.

4.4   INSTITUTIONAL BARRIERS

The bureaucratic nature of governments all around the world can act as a barrier to effective climate action in several ways. Problems stem from issues such as complex regulatory frameworks, decision-making processes, lack of coordination, limited flexibility, political influence, lack of accountability, resource constraints, and legal barriers.

Bureaucracies are often structured with multiple layers of governance and a complex hierarchy. Each level may require approvals and consultations, which can significantly delay the implementation of climate policies. The urgency of climate action is compromised by the time it takes for a policy to navigate through the various bureaucratic tiers.

The complex funding process of governments is a major problem. Initiating any climate-related project often involves securing funding, which can be a complex process involving numerous proposals, justifications, and approvals. The intricacy of this process can delay the release of funds and, hence, the execution of climate initiatives.

Regulatory rigidity is a common barrier to solving climate change problems. Governments operate under a set of established regulations and laws that are not always swiftly adaptable to the changing needs required for climate action. These regulations can be outdated and might not accommodate innovative technologies or approaches necessary for reducing emissions or promoting sustainability. In some countries, there may also be legal barriers to climate action, such as existing laws and regulations that favor fossil fuel industries.

The political nature of governments means they can be susceptible to influence by powerful lobby groups representing industries that stringent climate policies may negatively impact. Such groups can push for the continuation of the status quo or water down proposed climate action policies.

Interdepartmental competition is a common problem in governments. Different government departments may have competing priorities and vie for the same resources. This can lead to conflicts and a lack of unified direction in addressing climate issues, which require a coordinated approach across all sectors.

The political system in many countries where governments are elected for a given period means bureaucratic institutions often operate on short-term political cycles, focusing on immediate outcomes rather than the long-term strategies required for effective climate action. Consequently, there can be a lack of commitment to long-term goals that are critical for addressing climate change.

The sheer amount of paperwork, compliance checks, and procedural steps needed to move any initiative forward can be daunting. This administrative burden can impede quick action and discourage innovative approaches due to the fear of non-compliance or the risk of audit complications.

Effective climate action often requires specialized knowledge and expertise that may not be present within the existing bureaucratic framework. Many governments lack such specialized knowledge and expertise. Without the input of scientists, environmental economists, and other experts, policies may be less effective or even counterproductive.

Finally, there is inadequate coordination on a global level. Climate change is a worldwide issue that requires coordination not only within national governments but also between countries. Bureaucratic systems can struggle with such coordination, especially when it involves cross-border policies, shared resources, or international agreements.

To overcome these barriers, some governments are trying to streamline bureaucratic processes, engage with stakeholders more effectively, invest in capacity building within government agencies, and push for greater interdepartmental coordination. However, these changes require a cultural shift within the bureaucracy, a reevaluation of priorities, and often a redefinition of the roles and responsibilities of public institutions in the context of the global climate crisis.

4.5   SOCIAL AND CULTURAL BARRIERS

Social and cultural barriers are significant factors that can slow down action on climate change. These barriers can be deeply embedded in the values, beliefs, practices, and social norms of a society, influencing both individual and collective actions. It is essential to recognize that cultural beliefs and values are numerous, complex, and can change over time. Some of the common social and cultural barriers include values and beliefs, political and ideological polarization, religious beliefs, social norms and practices, economic growth, inequality and power dynamics, education and awareness, and resistance to change.

Values and Beliefs

In cultures that value material wealth and consumption, reducing consumption or changing consumption patterns to be more sustainable can be challenging. Some societies prioritize short-term gains over long-term sustainability and may resist changes that require immediate sacrifice for future benefit. In other societies, there is a displacement of responsibility. People may recognize the issue of climate change but feel that it is the responsibility of others (government, businesses, or other countries) to take action, leading to a lack of personal commitment to change. Still, others believe that environmental outcomes are beyond human control and can lead to inaction.

Political and Ideological Polarization

In some regions, climate change has become a partisan issue, where acknowledgment and action on climate change are aligned with political identity. In some societies, skepticism about climate change is prevalent, often fueled by misinformation and politicization of the issue. This can lead to a lack of support for policies aimed at mitigating climate change. Still, in many societies, traditional ways of life and skepticism towards new technologies or practices can delay the adoption of sustainable alternatives, like renewable energy or plant-based diets.

Religious Beliefs

Some religious beliefs may conflict with the concept of human-induced climate change or the need for environmental stewardship, potentially slowing down action. For example, religious or cultural beliefs that humans have dominion over nature can justify the exploitation of natural resources. Some religious beliefs that emphasize an impending end of the world may diminish the motivation to address long-term issues like climate change.

Social Norms and Practices

Established norms and practices that are heavily centered on consumption and waste are unsustainable. For example, owning large vehicles or taking frequent flights are at odds with sustainable practices. Dietary preferences, such as high meat consumption, which have a more significant environmental impact, can be hard to change due to cultural significance. Some of these high-status consumption patterns may be aspirational and signify success. They are difficult to change and can persist as a barrier to adopting more sustainable behaviors. In addition, people may be reluctant to adopt sustainable behaviors if they are not widespread or if they fear social disapproval. In many societies, traditional ways of life and skepticism towards new technologies or practices can delay the adoption of sustainable alternatives, like renewable energy or plant- based diets.

Economic Growth

Communities or countries that depend economically on fossil fuel extraction and use may resist transitioning to cleaner energy sources. These economies are focused on continuous growth, which may conflict with the need to reduce emissions and environmental footprints. In many regions, especially in developing countries, immediate economic survival outweighs long-term environmental sustainability. There may be cultural priorities placed on economic development and job creation that conflict with actions needed to address climate change.

Inequality and Power Dynamics

The unequal distribution of power and resources can result in marginalized groups having less of a voice in climate change discourse and decision-making, even though these groups are often the most affected by climate impacts. Concerns that climate policies may disproportionately affect certain groups can lead to resistance. Some cultures (particularly in developing countries) may resist change based on the belief that other nations historically responsible for greater emissions should lead the way.

Education and Awareness

Without proper education and awareness of climate change, individuals may not understand the urgency or importance of taking action. A lack of environmental education can lead to apathy or misunderstanding about the importance of climate action. The spread of misinformation can lead to confusion and hinder climate action. Lack of education and awareness can also lead to disconnection from nature. In many modern societies, people live in highly urbanized environments and may feel disconnected from the natural world. This disconnection can lead to a lack of awareness or concern for environmental issues.

Resistance to Change

Humans have a natural resistance to change, and shifting societal behaviors and values can be a slow process, especially when it involves widespread lifestyle changes. Likewise, institutions may resist change due to an ingrained culture that favors existing practices and norms. Societies may find it difficult to deviate from established pathways, such as energy systems and urban development patterns.

To overcome these barriers, strategies can include targeted education and communication campaigns that consider cultural contexts and efforts to integrate sustainable practices into cultural norms. It's also crucial for climate change action to be perceived as equitable and inclusive to garner widespread support across different social and cultural groups. Additionally, engaging with cultural and religious leaders to build supportive narratives around climate action can also be an effective way to catalyze change.

4.6   OPPOSITION FROM FRIENDS AND FAMILY

Friends and family help establish social norms, which are shared expectations about appropriate behavior. Opposition from friends and family can significantly affect individual and collective action on climate change in several ways.

Psychological and Cultural Impact

People are influenced by their social environment. If friends and family oppose climate change action, individuals may be less likely to adopt behaviors that mitigate climate change to conform with their group's norms. Similarly, when individuals feel unsupported by their close social circle in their environmental efforts, they may experience reduced motivation and possibly abandon these efforts. Cognitive dissonance also plays into the psychological impact. When one's beliefs are not aligned with those of their close social group, it can lead to cognitive dissonance, which might result in the individual changing their beliefs to align with the group rather than facing the discomfort of disagreement.

Cultural opposition within a social group can lead to a conflict of values, where environmental concerns are overshadowed by other priorities, such as economic growth or traditional practices that may be harmful to the environment. Social circles often share lifestyles and traditions. If these include environmentally unfriendly practices, they can perpetuate harmful behaviors across generations.

Behavioral Influence

Suppose a person's social circle is opposed to climate change action. In that case, this can directly influence the person's willingness to take action, such as reducing waste, conserving energy, or participating in activism. It can also influence their decision-making. Major life decisions, like buying a fuel-efficient vehicle or investing in renewable energy, can be swayed by the opinions of friends and family.

Socio-Political Impact

Friends and family can influence one's political views and actions. If a person's close circle is opposed to policies or candidates that support climate action, they may be less likely to support those policies or candidates themselves. Effective climate change action often requires collective effort. Opposition within one's immediate social group can lead to a lack of participation in larger group actions, which are crucial for substantial change.

Economic Influence

Friends and family affect consumer behavior. If they favor products or services that are not environmentally friendly, individuals may also be inclined to choose these options. For people who make investment decisions based on recommendations from their social circle, opposition to climate change could lead to supporting businesses that are not committed to sustainable practices.

Educational Influence

Friends and family are vital sources of information. If they disseminate misinformation about climate change or downplay its importance, they can contribute to a lack of awareness and misunderstanding about the issue. Educational influence on the next generation can be profound. Parents and close relatives play a significant role in shaping the values and beliefs of children. Opposition to climate action within the family can instill skepticism or indifference in the younger generation.

Overcoming Opposition

Engaging in constructive conversations and providing evidence-based information can sometimes sway friends and family to reconsider their stance. Engagement at the community level can also be important. Joining or forming groups with like-minded individuals can provide the social reinforcement needed to counteract opposition in one's closer social circles. Identifying with role models or leaders who advocate for climate change action can help individuals stand firm in their beliefs and actions despite opposition.

4.7   LEGAL AND SECURITY CHALLENGES

The urgency to address the escalating climate crisis has galvanized a global movement, compelling activists to raise their voices, mobilize citizens, and push for substantial policy changes. Yet, the path of climate activism is fraught with a multitude of legal and security challenges. From the streets of global cities to the corridors of international climate summits, activists confront a complex landscape of legal hurdles and security risks that both constrain and define their work. This Section examines the multifaceted legal and security challenges faced by climate activists and the implications for the broader movement seeking environmental justice and systemic change.

Legal Challenges

In a disconcerting trend, governments across the globe have been criminalizing protest activities. Laws that expand definitions of trespass, vandalism, and public nuisance are increasingly being used to arrest and prosecute activists. The legal jeopardy of participating in civil disobedience— once a cornerstone of peaceful protest—has escalated, often branding climate activists as lawbreakers rather than as citizens engaged in democratic expression. During protests or other climate-related activities, activists may be arrested and detained. Sometimes, the conditions of detention and the treatment of detainees can be harsh and violate their rights. Activists can face criminal charges for their activities, particularly in countries where dissent is not tolerated. Charges can range from minor offenses like loitering to more severe accusations like incitement or terrorism, depending on the country's legal framework and attitude towards activism.

Corporations and individuals have found a powerful tool to stifle climate activism through Strategic Lawsuits Against Public Participation (SLAPP). These suits, although often meritless, can entangle activists in costly and protracted legal battles, draining their resources and deterring others from speaking out for fear of legal retribution.

When activists face legal challenges, access to justice can be problematic. They may lack the financial resources to afford legal representation, and in some jurisdictions, the judiciary may not be independent or may be biased against activists.

Under the guise of national security, some nations have conflated environmental activism with terrorism or radical extremism. Anti-terrorism statutes and public order laws have been employed to justify the surveillance, arrest, and harsh treatment of climate activists, thereby chilling participation and expression.

Security Challenges

The digital age has equipped authorities and private entities with sophisticated surveillance tools, enabling them to monitor activists' communications and movements. This can include monitoring communications, tracking movements, or infiltrating activist groups. Such intrusions into privacy chill free speech and assembly, as activists must weigh the risk of being watched against the imperative of their cause.

In some cases, climate activists can be subjected to violence from authorities or opposing groups. This is particularly acute in certain regions where environmental defenders are targeted for their work against logging, mining, or agricultural interests. This poses grave risks to activists' safety and security.

Climate activism's reliance on digital platforms exposes it to cybersecurity risks. Activists may face hacking attempts, phishing, and other cyber-attacks aimed at disrupting their work or stealing sensitive information.

Activists who are well-known or have been arrested for their activities might face travel restrictions, either from their governments or from foreign governments, that may deny them entry.

4.8   LACK OF RESOURCES

Climate activists often operate on limited budgets and may lack the financial resources to run effective campaigns, conduct research, or organize events, which in turn can limit the reach and impact of their efforts. Here are some of the ways in which budget constraints can affect climate activism.

Activists often rely on public events, workshops, and educational campaigns to spread awareness about climate issues. Mobilizing activists and materials to various locations, whether for protests, educational campaigns, or conferences, requires funding. With limited budgets, the scope of physical mobility is reduced, which can affect the reach and impact of activism. It can also make it difficult to sustain long-term campaigns and maintain momentum, which will be detrimental to their efforts, given that climate activism often requires sustained effort over long periods.

When there is no financial support for activists, especially those from underprivileged backgrounds, there is a barrier to entry that can exclude passionate individuals who cannot afford to volunteer their time without compensation.

Limited budgets can also restrict the ability of activists to travel and participate in important events, such as conferences, meetings, and protests. This can limit their ability to connect with other activists, learn from experts, and build a broader movement for climate action. Budget constraints can also limit the ability of activists to build coalitions and partnerships with other organizations. This can hinder the ability to collaborate and amplify the impact of climate action.

In-depth research to support policy recommendations and the development of innovative solutions often require significant investment. Limited budgets mean these critical activities may be underfunded, affecting the quality and depth of information activists can provide.

Lobbying and policy advocacy often require sustained engagement with policymakers, which can include hosting events, preparing extensive policy documents, and following up on legislative processes. Limited budgets may result in less influence in these areas due to an inability to maintain a consistent presence.

Activism can sometimes lead to legal confrontations, particularly when challenging large corporations or government policies. Limited resources can make it difficult to sustain prolonged legal battles, which are often expensive.

Effective communication strategies, such as social media campaigns, video production, and website maintenance, are crucial for modern activism. Insufficient funds can hamper these efforts, limiting activists' ability to spread their message and engage with a broader audience.

4.9 ACTIVIST FATIGUE

In confronting the colossal specter of climate change, activists, scientists, and concerned citizens alike find themselves in a David-and-Goliath battle of existential proportions. The enormity of climate change manifests in both spatial and temporal dimensions. Spatially, the impacts are global; no corner of the Earth is untouched by the fingerprints of a changing climate, from the bleaching coral reefs of Australia to the melting ice sheets of Greenland. Temporally, the effects of climate change span across generations, making it a unique problem where the causes and consequences are decoupled by decades, if not centuries. This disconnect between actions and repercussions complicates the psychological grasp of the issue and dampens the immediacy of response.

Activists set out to tackle climate change with a sense of urgency and moral imperative, but the scale of the problem can quickly lead to a sense of insignificance and despair. Individual and even collective actions can seem like mere whispers against the roar of global industrial emissions, deforestation, and the inertia of societal and economic infrastructures resistant to change. The quantifiable goals, reducing carbon emissions to certain levels by specific dates, feel abstract when faced with the daily realities of policy negotiations, corporate lobbying, and the seductive comfort of status quo lifestyles.

This overwhelming scale can lead to what is known as "activist fatigue," a state of emotional exhaustion, cynicism, and a diminished sense of personal accomplishment. Activists who once vigorously campaigned for systemic change might find themselves questioning the impact of their efforts, suffering from burnout, and withdrawing from activities they once found meaningful. The ceaseless stream of dire predictions and data can turn the climate crisis into a leviathan that feeds on hope and energy, leaving behind a sense of helplessness.

Moreover, the relentless pace at which climate change must be addressed does not align well with human cognitive and emotional processing. The urgency dictated by science is at odds with the slow grind of policy change and the pace of cultural shifts. As activists push against these tides, the chronic stress of the "race against time" can lead to fatigue. They are not only fighting against climate change but also against the psychological toll of a problem that is too large, too complex, and too entrenched.

Furthermore, the interconnectedness of climate change with other societal issues, such as economic inequality, racial injustice, and global health, adds layers of complexity that can be daunting. For many activists, what starts as a campaign to reduce carbon footprints expands into a crusade against a hydra of interrelated problems, each as challenging as the next. The scale of climate change is not just a measure of carbon in the atmosphere; it is a reflection of the deeply woven patterns of human civilization.

However, understanding the nature of activist fatigue is the first step in mitigating it. Recognizing the signs of emotional exhaustion and taking proactive measures to address it is essential. Activists must find balance, pace themselves, and set realistic expectations to maintain their engagement over the long term. Collective action, support networks, and self-care strategies are crucial in sustaining the morale and effectiveness of those on the front lines of climate advocacy.

The key to overcoming the paralysis that can accompany the scale of the climate change problem may lie in reimagining the nature of action. Small, local, and incremental changes can build a mosaic of action that resonates on a global scale. Celebrating victories, no matter how minor they may seem in the grand scheme, can reignite the flames of hope and resilience. Grounding activism in the community can also distribute the weight of the problem, making it more manageable and fostering a shared sense of purpose.

5

RENEWABLE ENERGY AND SUSTAINABILITY

5.1 RENEWABLE ENERGY

Renewable energy refers to power derived from resources naturally replenished on a human timescale, such as sunlight, wind, rain, tides, waves, and geothermal heat. The increasing reliance on renewable energy is essential to address the pressing issues of climate change, environmental degradation, and depletion of fossil fuels. This Chapter explores the various types of renewable energy, their benefits, challenges, and the steps necessary to transition to a more sustainable energy future.

Types of Renewable Energy

Solar Energy is the most abundant form of renewable energy derived from the sun's rays. Solar power can be harnessed through photovoltaic cells that convert sunlight directly into electricity or solar thermal systems that use mirrors or lenses to concentrate sunlight and produce steam to drive turbines.

Wind energy is generated by harnessing the kinetic energy of moving air to turn turbines that produce electricity. The amount of wind energy produced depends on wind speed, air density, and blade length.

A third form of renewable energy is **hydropower**. This involves generating electricity by harnessing the energy of flowing or falling water. It is typically done by damming rivers to create reservoirs, where the released water turns turbines.

Fourth is **biomass energy** (or bioenergy). This energy is derived from organic materials like wood, crops, and waste. These materials can be burned directly for heat or converted into biofuels like ethanol and biodiesel.

Geothermal energy is a fifth form of renewable energy derived from the Earth's internal heat. This heat can be harnessed by drilling wells and using steam to drive turbines.

Finally, we have **tidal and wave energy** generated by harnessing the movement of water caused by tides and waves, respectively.

Benefits of Renewable Energy

The benefits of renewable energy sources are numerous and include:

(1)   Renewable energy sources, such as solar, wind, hydro, and geothermal power, produce little to no greenhouse gas emissions or air pollutants, reducing the environmental impact and mitigating climate change.

(2)   By harnessing renewable energy, we decrease our reliance on finite fossil fuels like coal, oil, and natural gas, promoting energy security and reducing the risk of energy price volatility.

(3)   Renewable energy contributes to a significant reduction in carbon dioxide and other greenhouse gas emissions, which are the primary drivers of global warming and climate change.

(4)   It can be harnessed locally, reducing dependence on imported energy resources and increasing a nation's energy self-sufficiency.

(5)   The renewable energy sector creates jobs in manufacturing, installation, maintenance, and research and development, stimulating economic growth and diversification.

(6) Many renewable technologies are highly efficient and can be integrated with energy storage solutions to provide reliable and consistent power.

(7) The energy sources are sustainable because they are continuously replenished, unlike finite fossil fuels, which are subject to depletion.

(8) The use of renewable energy reduces air and water pollution, leading to cleaner and healthier environments for communities near energy production facilities.

(9) Investing in renewable energy technologies drives innovation and development of new technologies, spurring economic growth and competitiveness.

(10) Distributed renewable energy sources, like solar panels on rooftops, can enhance grid reliability by reducing strain on centralized power systems and decreasing the risk of blackouts.

(11) Renewable energy projects often take place in rural areas, providing economic opportunities and additional income for farmers and landowners who lease their land for wind turbines or solar arrays.

(12) While there can be upfront costs in installing renewable energy systems, over time, they often result in lower energy bills due to free and abundant fuel sources.

(13) Renewable energy technologies can be deployed in remote or off-grid areas, extending access to electricity and improving the quality of life for underserved populations.

(14) A diverse energy mix that includes renewables increases resilience against supply disruptions and market fluctuations.

Challenges of Renewable Energy

The challenges of renewable energy adoption include:

(1) Many renewable energy sources, such as solar and wind, are intermittent and variable, depending on weather conditions and time of day, which can lead to issues with grid stability and reliability.

(2) Developing cost-effective and efficient energy storage solutions is crucial to store excess energy generated during peak production periods for use when renewable sources are not producing power.

(3) Integrating renewable energy into existing energy grids requires significant upgrades and investments to accommodate fluctuations and distributed generation.

(4) Some renewable technologies, like large solar or wind farms, require substantial land or space, which can lead to land-use conflicts and environmental impacts.

(5) The upfront costs of installing renewable energy systems, such as solar panels or wind turbines, can be significant, although they often pay off over time through reduced energy bills.

(6) Compared to fossil fuels, renewable energy sources often have lower energy density, which means more land or resources may be needed to generate the same amount of energy.

(7) The predictability of renewable energy generation can be challenging, making it difficult for energy providers to plan and ensure consistent power supply.

(8) Shifting to renewable energy may require significant infrastructure changes, particularly in the transportation sector for electric vehicles and charging infrastructure.

(9) Continued research and development are needed to improve the efficiency and cost-effectiveness of renewable technologies.

(10) The availability of renewable resources varies by location, and not all regions have the same access to consistent renewable energy sources.

(11) While renewable energy sources have lower environmental impacts compared to fossil fuels, some technologies, such as large hydroelectric dams, can have significant ecological consequences.

(12) Transitioning from existing fossil fuel-based infrastructure to renewable energy can be expensive and may involve economic disruptions for certain industries.

(13) Inconsistent or inadequate energy policies and regulations can hinder the growth and adoption of renewable energy technologies.

(14) The production and disposal of materials used in energy storage technologies, such as batteries, can have environmental and supply chain challenges.

(15) Communities may resist renewable energy projects due to concerns about aesthetics, noise, or potential impacts on property values.

Addressing these challenges requires a combination of technological innovation, supportive policies, investment in infrastructure, and public engagement to ensure a successful transition to a more sustainable energy future.

Steps for a Sustainable Energy Future

Listed below are some of the steps for a sustainable energy future.

(1) Investing in research and development can lead to technological advancements that make renewable energy more efficient and cost-effective.

(2) Governments can provide incentives and subsidies to encourage the adoption of renewable energy.

(3) Educating the public about the benefits of renewable energy can drive demand for cleaner energy sources.

(4) Updating and strengthening the energy infrastructure is essential for integrating renewable energy sources.

(5) Global cooperation is necessary to address the challenges of climate change and transition to a sustainable energy future.

In summary, the transition to renewable energy is crucial to address the environmental, social, and economic challenges our dependence on fossil fuels poses. While there are challenges to overcome, the benefits of renewable energy, including environmental protection, economic growth, and energy security, make it a promising solution for a sustainable future. It requires a collective effort from governments, businesses, and individuals to invest in clean energy, support innovation, and advocate for policies that promote sustainability. As technology advances and costs decrease, renewable energy will play an increasingly important role in meeting the world's energy needs.

5.2 CONSERVATION

Conservation, a term derived from the Latin word "conservare," meaning to protect or preserve, has become a cornerstone in the global effort to sustain our planet's natural resources and biodiversity. Conservation encompasses a broad range of actions aimed at safeguarding ecosystems, species, and natural habitats, with the ultimate goal of maintaining the Earth's biological diversity and ensuring the availability of resources for future generations.

Importance of Conservation

Biodiversity is essential for ecosystem stability and resilience. Diverse ecosystems are better equipped to withstand environmental changes and disturbances, such as climate change, pollution, and habitat destruction.

Humans rely on biodiversity for food, medicine, and raw materials. Conservation of ecosystems ensures the availability of these resources and supports human livelihoods.

Types of Conservation

There are two types of conservation: in-situ and ex-situ conservation. In-situ conservation involves protecting natural habitats to sustain wildlife populations. This can include establishing protected areas, such as national parks, wildlife sanctuaries, and marine reserves.

Ex-situ conservation involves preserving species outside their natural habitats, such as in zoos, botanical gardens, and seed banks. This approach is often used for critically endangered or extinct species in the wild.

Conservation Strategies

One crucial conservation strategy is to promote sustainable development practices that balance economic growth with environmental protection.

Second, engaging local communities in conservation efforts can lead to more successful outcomes, as communities are often the primary stakeholders in natural resources.

Finally, implementing and enforcing legislation and policies that protect biodiversity and regulate human activities that harm the environment is essential for conservation success.

Challenges in Conservation

A few critical challenges in conservation include habitat destruction, climate change, and pollution.

Habitat destruction due to deforestation, agriculture, and urbanization significantly threatens biodiversity.

Climate change alters ecosystems and affects species distributions, making conservation efforts more challenging.

Finally, pollution, including air and water pollution, can devastate ecosystems and species.

In conclusion, conservation is critical in safeguarding our planet's natural resources and biodiversity. The importance of conservation cannot be overstated, as it is vital for ecosystem stability, human well-being, and the survival of countless species. To effectively address the challenges of conservation efforts, a multifaceted approach that includes sustainable development, community involvement, and strong legislation is required. Committing to conservation can ensure a healthy and thriving planet for future generations.

5.3   ECOSYSTEM RESTORATION

Ecosystem restoration is the process of assisting the recovery of an ecosystem that has been degraded, damaged, or destroyed. This process aims to re-establish an ecosystem's structure, function, and dynamics, making it more resilient and sustainable in the long term. Below is a discussion of the importance of ecosystem restoration, its various methods, challenges faced, and the need for global cooperation in this process.

Importance of Ecosystem Restoration

Healthy ecosystems are rich in biodiversity and crucial for the survival of numerous species. Ecosystem restoration helps conserve biodiversity by providing suitable habitats for flora and fauna.

Ecosystems such as forests and wetlands regulate climate by acting as carbon sinks. Restoring these ecosystems helps mitigate the impacts of climate change by sequestering atmospheric carbon dioxide.

Ecosystems such as wetlands act as natural water filters, improving water quality by removing pollutants. They also play a vital role in maintaining water availability by regulating the water cycle.

Many communities depend on ecosystems for their livelihoods. For example, forests provide timber, non-timber forest products, and other resources that contribute to the economy.

Methods of Ecosystem Restoration

Damaged ecosystems can be restored in several ways. Active restoration involves direct intervention, such as planting native species, removing invasive species, and controlling erosion. Ecosystems can also be restored passively by ceasing harmful activities and allowing the ecosystem to recover independently. Certain instances may require the combination of active and passive restoration, where human intervention is minimal, and natural processes are relied upon to restore the ecosystem. Restoring the natural balance of an ecosystem may involve reintroducing keystone species, such as predators

Challenges in Ecosystem Restoration

Successful restoration requires a deep understanding of the ecosystem, including its species composition, soil characteristics, and hydrology. Lack of knowledge can lead to unsuccessful restoration attempts.

The impacts of climate change, such as shifts in temperature and precipitation patterns, can complicate restoration efforts.

Second, ecosystem restoration can be expensive, and securing funding and resources can be a significant challenge.

Third, there are also social and cultural challenges. Restoration projects must consider the needs and values of local communities. Failure to do so can result in conflicts and project failure.

Global Cooperation for Ecosystem Restoration

Ecosystem restoration is not confined to national borders. Many ecosystems span multiple countries, such as rivers and migratory bird routes. Therefore, global cooperation is essential for the success of ecosystem restoration. International initiatives such as the Bonn Challenge and the United Nations Decade on Ecosystem Restoration aim to restore millions of hectares of degraded land worldwide.

In conclusion, ecosystem restoration is a crucial process that benefits biodiversity, climate regulation, water quality, and human livelihoods. It involves various methods, including active and passive restoration, assisted natural regeneration, and rewilding. However, the process faces numerous challenges, including lack of knowledge, funding, social and cultural issues, and climate change. Global cooperation is essential to overcome these challenges and achieve successful ecosystem restoration. By working together, we can restore the health of our planet and secure a sustainable future for all.

5.4 SUSTAINABILITY

Sustainability, a term often used to describe a balance between the environment, social equity, and the economy, is crucial to ensuring the well-being of future generations. It aims to meet the needs of the present without compromising the ability of future generations to meet their own needs. Sustainability is interconnected with responsible management and stewardship of natural resources, social justice, and economic development.

Sustainability is often divided into three primary pillars: environmental sustainability, social sustainability, and economic sustainability.

The first is environmental sustainability. This involves the responsible management and conservation of the Earth's natural resources. The objective is to maintain the health of the planet and its ecosystems by reducing pollution, preserving biodiversity, and mitigating climate change. Examples of environmental sustainability practices include using renewable energy sources, such as solar and wind power, promoting energy efficiency, recycling, and the protection of natural habitats.

On the other hand, social sustainability focuses on improving individuals' and communities' quality of life and well-being. It includes promoting social equity, human rights, and access to education and healthcare. Social sustainability also involves the creation of inclusive and participative communities that value diversity and promote social cohesion. An example of social sustainability is providing equal access to educational opportunities, ensuring that all members of society have the tools they need to succeed and contribute positively to their communities.

The final pillar is economic sustainability. This involves the creation of a stable and resilient economy that provides opportunities for all members of society to thrive. It includes promoting fair trade practices, supporting local economies, and ensuring businesses operate ethically and responsibly. Economic sustainability also involves the implementation of policies that foster long- term economic growth and stability. An example of economic sustainability is investing in green industries that create jobs while addressing environmental challenges.

The three pillars of sustainability are interrelated and mutually reinforcing. For instance, protecting the environment can lead to improved health outcomes and economic opportunities. Similarly, promoting social equity can foster a sense of community and shared responsibility for the planet, which can drive sustainable economic development.

However, achieving sustainability is a complex and challenging task that requires the collective efforts of governments, businesses, and individuals. It requires a shift in mindset and values and the development of new technologies and practices that minimize harm to the planet and its inhabitants.

In conclusion, sustainability is a comprehensive approach that seeks to balance the needs of the environment, society, and the economy. It is a critical concept that can help ensure future generations' well-being by protecting the planet, promoting social equity, and fostering economic development. Achieving sustainability requires the collective efforts of all members of society and a shift in mindset and values towards a more responsible and sustainable way of living.

6

GLOBAL PERSPECTIVE

Climate change is a global issue that affects countries and regions differently. International perspectives on climate change vary depending on a nation's level of development, geographic location, and political priorities. The following is an overview of efforts and challenges related to climate change in different representative regions of the world.

6.1 NORTH AMERICA

The United States has experienced a shift in climate policy with changing administrations. While some states and cities have taken ambitious steps to combat climate change, federal efforts have varied. Challenges include political polarization and the need for a transition from fossil fuels.

Canada has committed to reducing greenhouse gas emissions and transitioning to a low-carbon economy. However, the country faces challenges related to its reliance on the oil and gas industry, particularly in Alberta.

6.2 EUROPE

The EU is a global leader in climate action, with ambitious targets to achieve carbon neutrality by 2050. The European Green Deal outlines a comprehensive plan for reducing emissions, promoting renewable energy, and fostering sustainable practices.

Nordic countries like Sweden, Norway, and Denmark are pioneers in renewable energy and sustainable transportation. They have set high renewable energy targets and implemented carbon pricing mechanisms.

6.3 ASIA

As the world's largest emitter of greenhouse gases, China has made significant strides in renewable energy adoption, electric vehicles, and afforestation efforts. However, it still faces challenges due to its heavy reliance on coal and rapid industrialization.

India is rapidly expanding its renewable energy capacity but faces challenges related to energy access for its large population. Balancing economic development with emissions reductions is a key challenge.

6.4 AFRICA

Many countries in Sub-Saharan Africa face the dual challenge of adapting to the impacts of climate change and reducing emissions. Lack of resources and infrastructure are significant barriers to climate action.

In North Africa, water scarcity and desertification are pressing issues in the countries of this region, which require sustainable solutions for agriculture and water management.

6.5 LATIN AMERICA

Brazil is home to the Amazon rainforest and has a crucial role in global climate efforts. Deforestation, driven by agricultural expansion, poses a significant challenge to climate action.

The small Central American nation of Costa Rica is a leader in sustainability, focusing on renewable energy, reforestation, and conservation efforts.

6.6 OCEANIA

Australia has faced criticism for its climate policies due to its continued reliance on coal and lack of ambitious emissions reduction targets. The country also experiences extreme weather events linked to climate change.

The low-lying Pacific Island nations are among the most vulnerable to rising sea levels. They advocate for urgent global action to mitigate climate change and support adaptation efforts.

6.7   MIDDLE EAST

The Oil-rich Gulf countries are investing in renewable energy projects but still rely heavily on fossil fuels. Water scarcity and extreme heat pose challenges to sustainability efforts.

Despite regional conflicts, Israel has made significant strides in water conservation and renewable energy development.

Efforts and challenges related to climate change vary widely across regions, and international cooperation is crucial to address this global issue effectively. The Paris Agreement, signed by almost every country in the world, represents a global commitment to limiting global warming and mitigating its impacts. However, the success of these efforts depends on individual and collective actions at the regional and national levels.

7

CLIMATE POLICY AND INTERNATIONAL AGREEMENTS

Addressing climate change is a complex global challenge requiring coordinated efforts from governments and international organizations. The roles of governments and international organizations in combating climate change include setting policies, providing funding and resources, conducting research, and facilitating cooperation among nations. Here are the specific vital roles of governments and international organizations in addressing climate change.

7.1   SETTING POLICIES

Climate change stands as a formidable global challenge that transcends borders, demanding collective action and coherent policy responses from both national governments and international organizations. The complexities of climate science, coupled with the socio-economic and political dimensions of climate-related policies, necessitate an integrated approach to mitigate the adverse effects of climate change and to foster sustainable development.

Governments

National governments hold the primary responsibility for developing and enforcing policies within their territories. Their role is multi-faceted, involving regulation, funding, and incentivization of practices that contribute to climate mitigation and adaptation.

Regulatory Frameworks: Governments enact laws and regulations that limit greenhouse gas emissions, promote renewable energy, and encourage energy efficiency. For instance, carbon pricing mechanisms, such as carbon taxes or cap-and-trade systems, are effective tools to internalize the cost of carbon emissions and incentivize low-carbon technologies.

Research and Innovation Funding: Investment in research and development is crucial for advancing climate science and technology. National policies that fund clean energy research can lead to breakthroughs in renewable energy, energy storage, and carbon capture and storage (CCS) technologies.

Public Awareness and Education: Governments also play a pivotal role in raising public awareness about the impacts of climate change. Education policies that integrate climate science into curricula can equip future generations with the knowledge to make sustainable choices.

International Organizations

International organizations act as platforms for cooperation, knowledge exchange, and policy coordination among nations. Their role is crucial in setting global standards and facilitating collective action. Organizations such as the UNFCCC provide a forum for countries to negotiate international agreements that set targets and establish frameworks for climate action, exemplified by the Paris Agreement. International financial institutions like the Green Climate Fund (GCF) and the Global Environment Facility (GEF) help mobilize financial resources to support climate projects, particularly in developing countries that are often the most vulnerable to climate impacts but the least equipped to deal with them. International bodies assist in capacity building by providing technical expertise and support to help countries develop and implement effective climate policies. They facilitate the transfer of technology and best practices across borders.

Collaboration and Challenges

The intertwined responsibilities of governments and international organizations showcase the necessity for collaboration. However, challenges such as differing national interests, economic disparities, and the need for equitable policy measures complicate the path forward.

National policies must align with international objectives without compromising a country's own development goals. This requires a delicate balance between self-interest and global responsibility.

Climate policies must be fair and just, acknowledging that while the impacts of climate change are universal, they are not evenly distributed. International policies should support mechanisms like technology transfer and climate finance that address these inequalities.

The effectiveness of international agreements depends on compliance and accountability mechanisms to ensure that countries adhere to their commitments. This necessitates transparent monitoring and reporting systems.

7.2   CLIMATE FINANCING

The battle against climate change is not just a scientific and technological challenge but also a financial one. The implementation of climate change mitigation and adaptation strategies requires substantial funding and resources, which are often beyond the capacity of individual actors or nations, especially in the developing world. Governments and international organizations play a pivotal role in mobilizing and directing financial resources to where they are most needed in the fight against climate change.

National Funding Mechanisms and International Contributions

Many national governments allocate funds within their budgets to support renewable energy projects, energy efficiency, sustainable transportation, and climate resilience initiatives. By investing in green infrastructure and technologies, governments can reduce carbon footprints and stimulate job creation in emerging sectors.

Governments often collaborate with private sector entities to finance climate-related projects. Through instruments like green bonds and climate finance funds, they leverage additional capital from private investors who are increasingly interested in sustainable and socially responsible investments.

Developed countries have committed to providing financial support to developing nations to help them cope with the impacts of climate change and transition to low-carbon economies. This includes direct bilateral aid, as well as contributions to international climate funds.

Coordination and Redistribution of Funds

Organizations such as the United Nations oversee international frameworks for climate finance, ensuring that funds are raised, managed, and distributed equitably. This involves setting up dedicated funds like the Green Climate Fund (GCF) that are specifically designed to assist countries in their climate change efforts.

International organizations also play a critical role in helping countries, especially those with limited resources, to develop the capacity to access and utilize climate finance effectively. They offer technical assistance and training programs, enabling governments to design projects and proposals that meet the funding criteria of various climate funds.

To ensure that climate funds are used effectively, international organizations establish rigorous reporting and accountability frameworks.

These mechanisms are designed to track the flow of funds, measure the outcomes of funded projects, and ensure that the money is having a tangible impact on combating climate change.

Challenges and Innovations in Climate Finance

Despite the growing pool of climate finance, there is still a significant gap between the funds available and the investment required to meet global climate targets. Governments and international bodies continue to seek innovative funding mechanisms to bridge this gap.

There is an ongoing challenge to ensure that climate finance is distributed equitably, reaching those who need it most. Low-income and climate-vulnerable countries often face barriers to accessing funds due to complex application processes and stringent requirements.

While public funds are essential, they are insufficient to meet the total costs of climate action. A critical role of governments and international organizations is to create an enabling environment that can unlock private-sector finance and investment in climate solutions.

7.3   RESEARCH

The complexities of climate change necessitate an in-depth understanding of its causes, consequences, and mitigation strategies. Scientific research is the bedrock of informed policy-making, and governments, alongside international organizations, play a critical role in spearheading and funding such research.

National Leaders in Climate Research and Innovation

Governments have the authority and resources to drive research agendas within their borders, laying the groundwork for advancements in climate science and related technologies.

National governments are significant sources of funding for climate change research, from basic climate science to applied research on mitigation and adaptation technologies. Governments support universities and public research institutions that conduct a substantial portion of climate- related research. These institutions often become hubs for international collaboration. By identifying priority research areas, governments can direct efforts toward the most pressing climate-related challenges, ensuring that research outputs have direct policy relevance.

Facilitators of Global Research Collaboration

International organizations provide a global platform for pooling knowledge, resources, and efforts to address climate change from a multinational perspective.

Bodies like the Intergovernmental Panel on Climate Change (IPCC) aggregate and assess scientific research to provide consensus on climate change, its impacts, and potential strategies for mitigation and adaptation.

International organizations often coordinate large-scale research programs that transcend national borders, such as those focusing on ocean health, polar ice monitoring, and global atmospheric studies. To ensure comparability and collaboration, international organizations set standards and facilitate data sharing among researchers and nations, accelerating the pace of innovation and discovery.

The Symbiotic Relationship and Challenges

The dynamic between national governments and international organizations in climate research is not merely additive but synergistic. They must navigate challenges that arise from such interdependent relationships to maximize the impact of their research efforts.

Governments may have national interests that do not always align perfectly with global research priorities. International organizations play a critical role in aligning these varied interests toward a common goal. A major challenge lies in the equitable distribution of funds for research, ensuring that developing countries also can conduct vital climate research.

While research is crucial, its benefits are only realized if findings are effectively disseminated and translated into policy and practice. Governments and organizations must work together to bridge the gap between research and implementation.

7.4   INTERNATIONAL COOPERATION

Climate change, by its very nature, is a global challenge that knows no borders, and its solutions are contingent on the cooperation and collective action of nations worldwide. Governments and international organizations are the linchpins in orchestrating this global response, providing the frameworks and incentives for collaboration among countries to tackle the multi-faceted dimensions of climate change.

National Actors on the Global Stage

As sovereign entities, governments are responsible for representing their nation's interests on the international stage, yet they must navigate the complex interplay of domestic agendas and global climate commitments.

Governments engage in diplomacy and bilateral agreements to advance climate cooperation, often leading to shared initiatives such as cross-border renewable energy projects or joint conservation efforts. By formulating and communicating national commitments, such as Nationally Determined Contributions (NDCs) under the Paris Agreement, governments contribute to the collective global target of limiting warming to well below 2 degrees Celsius. Nations take turns hosting international climate summits, where they can lead by example, galvanize action, and create momentum for global climate initiatives.

International Organizations

International organizations serve as the architects and mediators of the global governance structures needed for effective international climate cooperation. Organizations such as the United Nations Framework Convention on Climate Change (UNFCCC) provide the primary platform for dialogue, negotiation, and setting the global climate agenda. Through international treaties and agreements, these organizations establish frameworks for collective action, outlining the roles, responsibilities, and targets of participating nations.

International bodies facilitate cooperation by providing technical assistance and financial mechanisms to support climate projects, particularly in developing nations.

Synergizing Efforts and Overcoming Barriers

The relationship between national governments and international organizations in combating climate change is characterized by their efforts to synergize global climate action while confronting several inherent barriers. One of the most significant challenges is to reconcile individual national interests with the collective needs of the worldwide community. It requires a delicate balance and often involves compromise and negotiation.

Ensuring that cooperation frameworks account for the historical responsibilities of developed nations and the unique needs of developing countries is vital for equitable climate action.

Trust is the cornerstone of international cooperation, necessitating transparent communication and reporting systems to build and maintain confidence among nations in their collective endeavors.

8

CLIMATE INJUSTICE

8.1   DISPARITIES AT THE LOCAL, NATIONAL, AND INTERNATIONAL LEVELS

Climate injustice refers to the inequitable distribution of the burdens and effects of climate change, as well as the unequal involvement and recognition in the solutions and decision-making processes related to climate mitigation and adaptation. At the local, national, and international levels, climate injustice manifests in various forms, exacerbating existing inequalities and often hindering unified action against global warming.

Local Level Injustices

At the local level, climate injustice is seen in the differential impacts of climate change on various communities within a region or country. Low-income neighborhoods and communities of color often face greater exposure to environmental hazards, including pollution and extreme weather events. For instance, urban heat islands disproportionately affect poorer areas with less green space, leading to higher mortality rates during heat waves. Moreover, these communities typically have less access to resources needed for resilience and adaptation, such as emergency services, information, and funding. Local injustices are compounded by political and economic marginalization, where the voices of the most affected are frequently sidelined in planning and policy-making, resulting in a lack of appropriate measures to protect these communities and enhance their adaptive capacities. These injustices exacerbate existing social and economic inequalities, leading to a vicious cycle where the most affected are least able to respond.

National Injustices

Nationally, climate injustice can be observed in the uneven contribution to and varying degrees of suffering from the effects of climate change across different regions of a country. National climate injustice is also reflected in domestic policies that may favor certain regions or sectors over others, often influenced by economic interests that overlook the needs of vulnerable populations. This includes insufficient investment in rural areas or in sectors like agriculture, which are highly susceptible to climate impacts but often lack the necessary support for sustainable practices and innovation. Socially, marginalized groups within a country, such as women, children, indigenous peoples, and racial minorities, often face the highest risks from climate change due to existing inequalities.

International Injustices

Geographically, regions such as Sub-Saharan Africa, South Asia, and small island developing states are more vulnerable to climate-induced disasters despite contributing minimally to global emissions. Despite international agreements such as the Paris Agreement, there is an evident gap in the commitments and actions of developed and developing countries. Wealthier nations have greater access to technology, funds, and expertise to transition to low-carbon economies and implement adaptive strategies. In comparison, poorer nations struggle to secure financial and technical assistance for their climate actions. The principle of common but differentiated responsibilities (CBDR) is often not fully realized in international climate negotiations. CBDR, formalized in international law at the 1992 United Nations Conference on Environment and Development (UNCED) in Rio de Janeiro, establishes that all states are responsible for addressing global environmental destruction yet are not equally responsible. The principle balances, on the one hand, the need for all states to take responsibility for global environmental problems and, on the other hand, the need to recognize the vast differences in levels of economic development between states. Nevertheless, with less developed nations having less influence on the global stage, this frequently leads to agreements that do not fully address or compensate for the unequal burdens they bear. Unequal impacts and resilience hinder global development goals, as they often undo years of progress in poverty reduction and improved health outcomes. Furthermore, they pose significant challenges to social cohesion and global governance, as affected communities may be forced to migrate, leading to potential conflicts and increased stress on urban infrastructure and services.

8.2   UNEQUAL DISTRIBUTION OF CLIMATE IMPACTS

The stark inequality in the distribution of climate impacts raises urgent moral, economic, and political challenges that must be addressed if we are to achieve a just and sustainable future. This effort requires a holistic and collaborative approach considering social, economic, and environmental factors to ensure a more equitable and sustainable future for all.

Climate Justice Policies

At the local and national levels, climate justice should involve creating inclusive policies that prioritize the needs of the most vulnerable and establish mechanisms for community-level participation in decision-making. The policies should include transparent and equitable climate governance that allocates resources and support reasonably across different regions and communities, ensuring that no group bears an undue burden of climate impacts. They should recognize the unique challenges faced by women in climate-affected areas and promote gender equality in climate adaptation and mitigation efforts. Government efforts should include raising public awareness about the unequal impacts of climate. This is one area where grassroots movements and civil society organizations can play a pivotal role. Policies should provide for educating and empowering vulnerable communities to understand and respond to climate change, building their capacity to adapt to changing conditions and engage in sustainable practices.

Mitigation Efforts

Addressing unequal climate impacts requires a significant emphasis on mitigation efforts to reduce overall greenhouse gas emissions. Aggressive mitigation efforts should be prioritized to reduce greenhouse gas emissions. This effort includes transitioning to renewable energy sources, increasing energy efficiency, and implementing carbon pricing mechanisms. Investments in green technologies and industries can create jobs and economic opportunities in vulnerable regions, helping to alleviate poverty and build resilience. Developed countries must lead the way in cutting emissions and transitioning to renewable energy sources, reflecting their historical responsibility and current capabilities. By preventing the worst impacts of climate change, we can reduce the burden on vulnerable communities. These efforts should also be coupled with support for developing nations to leapfrog to cleaner technologies and build resilient infrastructure. Governments should invest in research and data collection to better understand the localized impacts of climate change. This information can inform targeted interventions and policies.

Adaptation and Resilience Building

Adaptation strategies must be prioritized to shield vulnerable communities from the worst impacts of climate change. This includes investing in resilient infrastructure, particularly in areas most susceptible to climate impacts, implementing early warning systems, and enhancing agricultural practices to withstand changing conditions. Importantly, adaptation efforts must be locally informed and empower communities to shape the responses to their unique vulnerabilities.

Financing Climate Action

A significant increase in climate financing is imperative to address the unequal impacts of climate change. This includes fulfilling the commitments made by developed countries to mobilize $100 billion per year to support climate action in developing countries. Furthermore, innovative financing mechanisms such as green bonds and climate risk insurance can provide additional resources to fund mitigation and adaptation initiatives.

9
CLIMATE REFUGEES

Climate change significantly impacts displacement and migration patterns worldwide, and it is estimated that millions of people could be displaced in the coming decades. As global temperatures rise and extreme weather events become more frequent and severe, people are forced to leave their homes and seek refuge elsewhere. The following are some key ways climate change influences displacement and migration.

9.1   SEA-LEVEL RISE

Rising sea levels are one of climate change's most visible and immediate impacts. Coastal communities are increasingly vulnerable to flooding and coastal erosion. As sea levels continue to rise, many low-lying areas and small island nations risk becoming uninhabitable, displacing millions of people.

The low-lying Pacific island nations of Kiribati, the Marshall Islands, and Tuvalu face the threat of disappearing entirely due to rising sea levels. Some residents have already migrated to other countries, and entire populations may need to be relocated. For instance, in 2015, a family from Tuvalu was granted residence in New Zealand after claiming to be climate refugees. Kiribati bought land in Fiji in 2014 as a potential relocation site for its citizens if sea levels continue to rise.

Bangladesh is one of the most vulnerable countries to rising sea levels due to its low-lying geography. Nearly one-third of the country is less than five feet above sea level. This has displaced thousands of people from their homes, pushing them to migrate to urban areas like Dhaka. Estimates show that by 2050, around 15 million people could be displaced by rising sea levels. Erosion, salinization of agricultural land, and frequent cyclones further strain the ability of people to live in the country's coastal regions.

In Alaska and other Arctic regions, the impacts of climate change are particularly acute, with temperatures rising at more than twice the global average. This has resulted in the melting of sea ice and permafrost, as well as coastal erosion, threatening the livelihoods of indigenous communities and forcing some to relocate. Some indigenous communities, like those in the village of Newtok, are relocating due to the changing landscape threatening their homes and way of life.

9.2 EXTREME WEATHER EVENTS

Climate change leads to more frequent and intense extreme weather events, such as hurricanes, cyclones, droughts, and wildfires. These events can destroy homes, infrastructure, and agricultural land, forcing people to flee their communities for safer places to live.

Central America's Dry Corridor (parts of Guatemala, Honduras, El Salvador, and Nicaragua) has been experiencing irregular rainfall patterns and prolonged drought periods, affecting food security. Many small-scale farmers, unable to grow crops due to these conditions, have considered migrating in search of better opportunities.

Saltwater intrusion in the low-lying Pacific island nations of Kiribati, the Marshall Islands, and Tuvalu contaminates drinking water, affects food crops, and destroys homes, thus threatening their health and livelihoods. This has caused many families to relocate within and outside the islands.

9.3 DESERTIFICATION, DROUGHT, AND FOOD INSECURITY

Changes in rainfall patterns and increased temperatures have led to the spread of deserts and more frequent and severe droughts. This makes it harder for people to grow food and access fresh water, which can force them to leave their homes in search of better living conditions.

The Sahel, a transitional zone between the Sahara desert in the north and the Sudanian Savanna in the south, includes parts of Senegal, Mauritania, Mali, Burkina Faso, Niger, Nigeria, Chad, Sudan, and Eritrea. Droughts, overgrazing, and deforestation have led to desertification, reducing the available land for agriculture and grazing. Lake Chad, for example, has shrunk by about 90% since the 1960s, primarily due to climate change, overuse of water resources, and poor water management. This has resulted in water scarcity and food insecurity for the millions of people who rely on the lake for agriculture, fishing, and livestock. As a result, many have been forced to migrate for better livelihoods.

In Central America, particularly in the "Dry Corridor" that spans parts of Guatemala, Honduras, El Salvador, and Nicaragua, prolonged droughts and extreme weather events have undermined food security and livelihoods, contributing to migration to other countries, including USA.

9.4   CONFLICT AND RESOURCE SCARCITY

Climate change can exacerbate existing conflicts and create new ones by increasing competition for scarce resources, such as water and arable land. Climate-induced resource scarcity can lead to displacement and further instability in regions with a history of conflict.

Between 2006 and 2011, Syria experienced a severe drought that led to crop failure, loss of livestock, and the depletion of water resources, causing the displacement of millions of rural dwellers to urban areas and intensifying socioeconomic tensions. Some analysts argue that this water scarcity and food insecurity contributed to the civil unrest that ultimately led to the Syrian Civil War, which has displaced millions of people within the country and neighboring nations.

The loss of livelihoods and resources in Lake Chad has fueled conflict in the region, notably with the rise of the terrorist group Boko Haram, contributing to the displacement of millions of people within and across national borders.

9.5   HEALTH IMPACTS

Climate change can affect public health by causing the spread of diseases and heat-related illnesses. In some cases, deteriorating health conditions can force people to migrate for better healthcare and living conditions.

In India and Pakistan, extreme heatwaves in recent years have caused thousands of deaths, especially among vulnerable populations like the elderly. As temperatures continue to rise and heatwaves become more frequent and intense, they can exacerbate existing health issues, push people out of non-air-conditioned urban areas, and reduce agricultural productivity, leading to migration.

Rising sea levels in Bangladesh are causing saltwater intrusion into freshwater sources, impacting drinking water and agricultural lands. This, coupled with the increasing number of cyclones, threatens health through salt-induced hypertension and reduced agricultural yields, pushing people to migrate from coastal regions to cities or neighboring countries.

In Latin America, as temperatures rise and rainfall patterns shift, the habitat range of mosquitoes carrying diseases like dengue, Zika, chikungunya, and malaria expands. This has been observed in parts of Central and South America, where changing climate patterns have led to the spread of these diseases to previously unaffected areas. This can strain healthcare systems, result in morbidity and mortality, and become a factor in people's decisions to migrate to regions less affected by these diseases.

9.6   CONSEQUENCES OF CLIMATE REFUGEE MIGRATION

There are several consequences of climate refugee migration. First, as people are displaced from rural areas due to climate impacts, they often move to urban centers in search of better opportunities and services. This can lead to rapid urbanization and strain on urban infrastructure.

Second, in some cases, the impacts of climate change can lead to cross-border migration that can strain countries receiving climate refugees. The international legal frameworks for recognizing and protecting climate refugees are still evolving.

Third, climate change does not affect everyone equally. Vulnerable populations, such as low-income communities, indigenous peoples, and marginalized groups, often bear the brunt of its effects. These people may have fewer resources to adapt and are more likely to be displaced.

Finally, governments and international organizations increasingly recognize the link between climate change and displacement. Efforts are underway to develop policies and strategies to address displaced populations' needs and mitigate the root causes of climate-induced migration through emissions reduction and climate adaptation.

In conclusion, climate change is a significant driver of displacement and migration, and its impacts are expected to grow in the coming years. Efforts to address this issue require strategies, including reducing greenhouse gas emissions, building resilience in vulnerable communities, and developing effective policies and mechanisms to assist and protect those displaced due to climate change.

10

CLIMATE COMMUNICATION

Effective science communication and storytelling are essential tools for raising awareness about climate change and inspiring action. Climate change is a complex and urgent global issue that requires the engagement of diverse audiences, including policymakers, the general public, and businesses. Here are some key strategies and principles for effective climate change communication:

10.1 KEY STRATEGIES AND PRINCIPLES

Know Your Audience: Tailor your message to your audience's interests, values, and level of understanding. Climate change affects various sectors and demographics differently, so consider your target audience's concerns and motivations. Get a good handle on the following about your audience:

(1) Demographics: Understand the age, gender, occupation, education level, and cultural background of your audience.

(2) Beliefs and Values: Tailor your message to resonate with their beliefs and values.

(3) Existing Knowledge: Begin from where they are in terms of knowledge about climate change.

Use Clear and Simple Language: Avoid jargon and technical terms that may confuse people. Use plain language and explain scientific concepts in simple terms. Graphics, visuals, and metaphors can also help convey complex ideas.

Tell Compelling Stories: Stories have the power to humanize climate change. Share personal stories, anecdotes, or case studies that illustrate the real-world impacts of climate change on individuals, communities, or ecosystems.

Stories create emotional connections and resonate with people.

Highlight Local Relevance: Emphasize how climate change affects the local community or region of your audience. People are more likely to act when they see a direct connection between the issue and their lives.

(1) Personal Stories: Share stories of individuals affected by climate change.

(2) Personal Responsibility: Explain how individual actions can contribute to or mitigate climate change.

Provide Solutions: Present actionable solutions and pathways for positive change. Show how individuals and communities can contribute to mitigating climate change and adapting to its effects. People are more likely to engage when they feel they can make a difference.

Utilize Varied Media Platforms:

(1) Social Media: Use platforms popular with your target audience.

(2) Documentaries & Films: Visual storytelling can be very impactful.

(3) Podcasts & Radio: Reach audiences during commutes or downtime.

(4) Local Workshops & Community Events: Engage directly with communities

Use Visuals and Data: Visuals such as maps, charts, and infographics can make data more accessible and compelling. Visual representations of the impact of climate change, such as before-and-after photos, can be particularly effective.

Address Misconceptions: Acknowledge common misconceptions or skepticism about climate change and provide clear, evidence-based explanations to counter them. Avoid confrontational approaches, as they can be counterproductive.

Leverage Emotions, But Carefully

(1) Hope Over Fear: While conveying the seriousness of the issue is okay, it's also essential to inspire hope. A message purely of doom and gloom can lead to feelings of helplessness.

(2) Empathy: Relate to people's daily experiences and challenges.

Build Trust: Establish credibility by citing reputable sources and experts. Transparency and honesty are crucial for building trust with your audience. Address uncertainties in the science while emphasizing the consensus among experts.

Foster Two-Way Communication: Encourage dialogue and engagement with your audience. Listen to their concerns, answer questions, and provide opportunities for feedback and discussion. Building a sense of community and shared responsibility is vital.

Collaborate and Amplify: Collaborate with organizations, influencers, and leaders who share your goals. Amplify your message through partnerships and social media to reach a broader audience.

Maintain Consistency: Consistency in messaging is essential for building a long-term understanding of climate change. Repeated and coherent messaging can reinforce the urgency of the issue.

Celebrate Small Victories: Celebrating small achievements in tackling climate change can motivate individuals and communities to take more significant steps.

Visuals Are Key:

(1) Photographs: Powerful images can evoke emotion – for instance, polar bears on thin ice or drought-stricken areas.

(2) Infographics: These can simplify complex data and make it more digestible.

(3) Maps: Show geographic changes over time due to climate impacts.

Repetition and Consistency: Repetition reinforces memory. Use consistent messaging across platforms and repeat key points.

Use Art and Culture: Integrate music, literature, and art to convey the message of climate change in diverse and compelling ways.

Engage in Two-Way Communication: This entails encouraging feedback, questions, and discussions, and addressing doubts and misinformation promptly and respectfully.

Keep Updated with the Latest Science: The field of climate science is continuously evolving. Stay informed to communicate the most accurate information.

Collaborate with Influencers: Partner with well-known personalities or experts who can lend credibility and broader reach to your message.

End with a Call-to-Action: Clearly state what you want the audience to do after they've received your message – whether it's a lifestyle change, supporting a policy, or joining a climate action group.

Remember, the goal isn't just to inform but to inspire action. Tailoring your communication approach based on your audience and context can make your message more impactful. Effective science communication and storytelling require ongoing efforts and adaptation to the evolving understanding of climate change. These strategies can help raise awareness, inspire action, and contribute to a more sustainable and resilient future.

10.2 MEDIA, ADVOCACY CAMPAIGNS & PUBLIC DISCOURSE

These play crucial roles in climate change communication by shaping public awareness, attitudes, and policy decisions. Here's an exploration of their roles in more detail:

The Media

The media plays a crucial role in climate change communication by disseminating scientific information, raising awareness, and influencing public opinion and policy. Here are some of the critical roles played by the media in this area:

(1) Traditional media outlets (TV, radio, newspapers) and digital platforms (websites, social media) are essential in conveying information about climate change to the public. They report on scientific findings, climate events, and policy developments.

(2) Media outlets frame climate change stories, which can influence public perception. For example, they may focus on the scientific consensus, the urgency of action, or the economic costs, shaping how people interpret the issue. Framing climate change as a human-caused problem requiring urgent action can motivate people to reduce their carbon footprint or support climate-friendly policies.

(3) Media can set the public agenda by determining which topics receive attention. By covering climate change prominently, they can encourage public and political discussions on the topic. If climate change is consistently covered as a pressing issue, it can increase public awareness and concern.

(4) Visual media, such as photographs and videos, can make climate change impacts more tangible and emotionally resonant, thus engaging the public on a deeper level.

(5) Some media outlets and journalists actively advocate for action on climate change and support campaigns aimed at reducing greenhouse gas emissions and promoting sustainable practices.

(6) Media can raise awareness about the impacts of climate change on different aspects of society and the environment. This includes highlighting the effects of extreme weather events, rising sea levels, and other climate-related issues.

(7) Media shapes public opinion and can influence people's behaviors and attitudes towards climate change. For example, media campaigns can encourage people to adopt more

sustainable lifestyles, such as reducing energy consumption, using public transportation, or supporting renewable energy.

(8)   Media may selectively report climate change, focusing on extreme events or controversies, which can skew public perception. Sensationalism and attention-grabbing headlines can sometimes lead to a distorted view of the issue.

(9)   Striking a balance between providing different perspectives and maintaining objectivity is challenging in climate reporting. Giving equal time to climate deniers can create a false sense of controversy when there is a scientific consensus on the issue.

(10)   Fact-checking organizations are vital in scrutinizing media reports for inaccuracies and misinformation. They correct false claims and provide accurate information to the public.

(11)   News outlets can consult climate scientists and experts to ensure accurate reporting. Having scientists review articles before publication can help prevent the spread of misinformation.

(12)   Media organizations can be transparent about their funding sources and affiliations, as this can influence their editorial stance. Independent and non-partisan reporting is essential for accurate climate coverage.

(13)   Many media organizations engage in educational efforts to increase public understanding of climate change. This may include explanatory articles, infographics, and interviews with experts.

(14)   Journalists and media professionals can undergo climate communication training to understand better the science, nuances, and complexities of climate change. This can help them convey accurate information effectively.

(15)   Engaging with the audience and addressing their questions and

concerns can build trust. Encouraging critical thinking and providing resources for further research can empower the public to discern accurate information from misinformation.

(16) Reporting on climate change can benefit from diverse sources, including local voices, impacted communities, and experts from various fields. This can provide a more holistic understanding of the issue.

(17) Combatting misinformation and promoting accurate climate reporting requires a sustained effort. Media organizations should commit to ongoing coverage and not just focus on sporadic events or developments.

(18) Media can influence policy by shaping public opinion and pressuring policymakers to take action on climate change. Investigative journalism can also expose the lack of action or inadequate policies addressing climate change.

(19) The media has an educational role in informing the public about climate change and what they can do to mitigate its effects. This can include providing practical tips on reducing carbon emissions and explaining the science behind climate change.

In summary, the media plays a crucial role in communicating climate change and shaping public opinion by disseminating information, framing the narrative, raising awareness, influencing public opinion and behavior, advocating for action, influencing policy, and providing education.

Advocacy Campaigns

Advocacy campaigns are crucial in climate change communication by raising awareness, educating the public, and driving action to address climate change. Here are some key ways advocacy campaigns contribute to climate change communication.

Advocacy campaigns help bring climate change to the forefront of public consciousness. They use various communication strategies, such as social media campaigns, public events, and media coverage, to highlight the impacts of climate change and the need for urgent action.

They provide valuable information and resources to inform the public about the causes and consequences of climate change. They often use scientific research and data to explain complex climate-related concepts in simple and accessible language.

Advocacy campaigns are critical in mobilizing public support for climate action. They organize events, rallies, and other activities that unite people to call for more robust climate policies and actions. They also encourage individuals to take personal steps to reduce their carbon footprint. They often use social media to create online movements and raise awareness.

Advocacy campaigns engage policymakers and stakeholders to push for more robust climate policies and regulations. They often collaborate with other organizations, experts, and communities to build a strong case for action.

Advocacy campaigns foster community and solidarity among people concerned about climate change. They provide a platform for people to come together, share their experiences, and support each other in their efforts to address climate change.

Finally, advocacy campaigns help amplify the voices of marginalized and vulnerable communities disproportionately affected by climate change. They ensure that these communities are represented in climate conversations and that their needs and perspectives are considered in climate policies and actions.

Overall, advocacy campaigns are a crucial component of climate change communication as they help to increase public awareness, education, and engagement on the issue, ultimately driving collective action to address the global challenge of climate change.

Public Discourse

Public discourse plays a crucial role in climate change communication by serving as a platform where different stakeholders, including policymakers, scientists, communities, and the general public, can engage in dialogue and exchange information, perspectives, and solutions.

The following are some of the key roles of public discourse in climate change communication:

(1) Public discourse helps to raise awareness about the causes, impacts, and solutions to climate change and can contribute to a better-informed public.

(2) Public discourse allows for exchanging scientific knowledge and traditional wisdom, leading to a more comprehensive understanding of climate change and its impacts.

(3) Through dialogue and discussion, public discourse can help build consensus on the need for action and the most effective strategies to address climate change.

(4) Public discourse can be a powerful tool for advocacy and mobilization, empowering communities and individuals to take action to mitigate and adapt to climate change.

(5) Public discourse can influence policy by providing a platform for stakeholders to voice their concerns, share their ideas, and advocate for specific policies or actions.

(6) Public discourse can hold policymakers, industries, and other stakeholders accountable for their actions and commitments related to climate change.

(7) Public discourse can foster stakeholder collaboration, leading to more effective and coordinated efforts to address climate

change.

(8) Public discourse can shape the framing of climate change. Stakeholders and interest groups may use discourse to advocate for their preferred policies or perspectives.

(9) Public discourse can influence social norms, making sustainable behaviors more socially acceptable and encouraging individuals to adopt climate-friendly actions.

(10) Public discourse can create a feedback loop with media and advocacy campaigns. As public discussions evolve, they can influence how media outlets cover climate change and how advocacy groups tailor their messaging.

Effective climate change communication relies on a symbiotic relationship among media, advocacy campaigns, and public discourse. The media disseminates information and frames issues, advocacy campaigns mobilize and engage the public, and public discourse shapes the broader conversation and policy landscape. Successful communication strategies often involve combining these elements to engage diverse audiences and drive action on climate change.

11

PERSONAL SACRIFICES AND CHALLENGES

Transitioning from a climate change denier to a climate change crusader can be a significant personal journey, marked by various sacrifices and challenges. This Chapter highlights some of the sacrifices and personal challenges someone like Prof. Jamie Cole might face in this transformation.

Changing Beliefs: The most fundamental sacrifice is the willingness to change one's deeply held beliefs. Admitting that climate change is real and a pressing issue requires a shift in mindset, which can be emotionally challenging. The individual may experience grief, guilt, or anxiety as they grapple with the implications of their previous beliefs and the urgent need for action.

Social and Professional Isolation: Prof. Jamie Cole faced isolation from friends, family, and professional networks that still hold climate-denying views. This can be a significant personal sacrifice, as these relationships may have been a fundamental part of the individual's support system.

Personal Relationships: Confronting climate change denial might lead to conflicts within personal relationships, particularly if loved ones deny the reality of climate change. This can strain bonds and even result in estrangement. Such was the case for Prof. Cole, whose marriage ended in divorce.

Career Implications: Depending on one's profession, advocating for climate action might be met with resistance or even backlash. For example, someone working in an industry closely tied to fossil fuels may face professional challenges or career changes.

Reputation Damage: The individual may have made public statements or taken actions in the past that denied climate change. As they shift their stance, they may face criticism or accusations of hypocrisy from climate activists and deniers. This can harm their reputation and credibility.

Personal Lifestyle Changes: Advocating for climate action often necessitates changes in one's lifestyle, such as reducing carbon emissions by using public transportation, eating a plant-based diet, or reducing energy consumption. These changes can be challenging and may require sacrifices.

Time and Energy Commitment: Becoming an effective advocate for climate action can be time- consuming and emotionally draining. It may require attending rallies, participating in environmental campaigns, or engaging in public speaking or education efforts.

Economic Consequences: If the individual's livelihood is connected to industries or practices contributing to climate change, they may face economic consequences as they transition to more sustainable alternatives. This could include job loss or financial instability.

Financial Sacrifices: In some cases, advocating for climate action might require financial sacrifices, such as investing in renewable energy sources or supporting environmental organizations.

Dealing with Denialist Backlash: Transitioning from being a climate change denier to a crusader can sometimes make one a target for backlash from the denialist community, which may include harassment, threats, or cyberbullying.

Learning Curve: The individual will need to educate themselves on the science of climate change, as well as the various solutions and policies that are being proposed or implemented to address it. This can be a time-consuming and potentially overwhelming process.

Public Criticism and Backlash: The individual may face criticism and backlash from both sides of the debate. Climate activists may be skeptical of their sincerity or question their motives, while former allies in the climate denial movement may see them as traitors.

Moral and Ethical Struggles: The individual may struggle with feelings of guilt or shame for their past beliefs and actions. They may also grapple with the moral and ethical implications of their previous denial and the harm it may have caused.

Activism Fatigue: Engaging in climate activism can be physically and emotionally demanding. Constant advocacy, attending protests, or participating in environmental campaigns can lead to burnout and exhaustion.

Uncertainty: Climate action can be frustrating due to the slow pace of change and the uncertain outcome. Crusaders may grapple with feelings of hopelessness and doubt about whether their efforts will make a difference.

Despite these challenges and sacrifices, many individuals who once denied climate change have become vocal advocates for climate action. Their unique perspective and journey can be powerful tools in convincing others to take the issue seriously and take action to mitigate the impacts of climate change.

12

TECHNOLOGICAL INNOVATIONS

Climate change is a complex and multifaceted challenge that requires innovative technologies and solutions to mitigate its effects. Researchers, scientists, and engineers worldwide are working on various approaches to address this problem. Here are some of the most promising technologies and solutions being developed to mitigate climate change:

12.1   RENEWABLE ENERGY TECHNOLOGIES

Renewable energy technologies harness energy from sources that are naturally replenished on a human timescale. These technologies provide alternatives to conventional fossil fuels and can reduce greenhouse gas emissions. Here's a rundown of some major renewable energy technologies

Harnessing Solar Power

The quest for sustainable and clean energy sources has long been at the forefront of environmental and economic agendas worldwide. Amidst the myriad of renewable energy technologies, solar power shines with unparalleled promise. Solar energy, harnessed from the inexhaustible radiation of the sun, stands as a beacon of hope in a world grappling with the harmful effects of fossil fuel consumption and the pressing need to decarbonize energy systems.

The Essence of Solar Energy: Solar energy derives from the sun's radiation. It can be converted directly into electricity using photovoltaics (PV) or indirectly with concentrating solar power (CSP), which uses lenses or mirrors to focus sunlight into a small area. The fundamental allure of solar energy lies in its abundance and ubiquity; the sun bathes the Earth with enough energy in one hour to meet global energy needs for an entire year. This inexhaustible supply, coupled with strides in solar technology, makes solar energy a cornerstone of renewable energy strategies.

Technological Advancements: Technological innovation in solar energy has been robust and transformative. The efficiency of solar panels has significantly increased while manufacturing costs have plummeted, making solar energy more accessible than ever. Advances in materials science have led to the development of organic photovoltaics and perovskite solar cells, pushing the boundaries of efficiency and versatility. Moreover, the integration of solar with energy storage systems, like batteries, and the development of smart grids have enhanced the reliability and utility of solar energy, facilitating its integration into existing energy infrastructures.

Benefits of Solar Energy: The adoption of solar energy offers myriad benefits that transcend environmental impact. On the environmental front, solar power is clean, producing no greenhouse gases during operation, thus mitigating climate change and reducing air pollution. Economically, it presents opportunities for growth in the green job sector and fosters energy independence for countries relying on imported fossil fuels. Additionally, solar installations can be deployed rapidly and scaled modularly, from small, off-grid rural systems to large utility-scale plants, making it a versatile solution for diverse energy needs.

Challenges and Solutions: Despite the promising trajectory, the expansion of solar energy faces several challenges. Energy production is intermittent, with variability in solar radiation due to weather and diurnal cycles. However, innovations in energy storage and grid management are addressing these reliability concerns. Another challenge is the environmental impact of solar panel production and disposal. The solution lies in developing recycling methods and embracing circular economy principles to manage the lifecycle of solar panels sustainably.

The Future of Solar Energy: The potential of solar energy as a cornerstone of our future energy mix is undeniable. International collaborations, such as the International Solar Alliance, exemplify the global commitment to harnessing solar energy. Governments are increasingly incentivizing solar power through subsidies and policy frameworks, propelling the industry forward. As we advance, the emphasis on research and development will continue to reduce costs and overcome technical barriers, fostering the integration of solar energy into a diversified, renewable energy portfolio.

Wind Energy

Wind energy, as a renewable energy technology, stands out as one of the most rapidly evolving and increasingly significant sources of clean power in the world's energy portfolio. Its principle is simple yet powerful: harness the kinetic energy of wind and convert it into electricity. This ancient practice, once used to sail ships and mill grain, is now refined by modern technology to generate electricity on a scale that can power entire communities.

Understanding Wind Energy: Wind energy relies on air currents caused by the uneven heating of the Earth's surface by the sun. Wind turbines, both onshore and offshore, capture this energy. As the wind blows, it passes over the blades of the turbine, causing them to turn. This rotational motion is transferred to a generator, which then produces electricity.

Technological Advancements: The technology behind wind turbines has matured significantly, leading to larger, more efficient, and more durable designs that are capable of generating substantial amounts of electricity even at low wind speeds and are cost-effective. Turbine blades are now designed with advanced aerodynamics and materials to capture wind more efficiently. The scale of turbines has also increased, allowing for greater energy capture; the newest turbines are so large that their rotors can sweep an area larger than the

London Eye. Moreover, improvements in forecasting technologies enable better prediction of wind patterns, optimizing the operation of wind farms and integrating their energy production into the power grid.

Benefits of Wind Energy: The benefits of wind energy are substantial. Environmentally, wind power is one of the cleanest sources of energy because it generates electricity without emitting greenhouse gases or pollutants. Economically, it provides a hedge against the volatility of fossil fuel prices and creates numerous jobs in manufacturing, installation, maintenance, and supporting services. Socially, wind energy projects often contribute to local community development through land lease payments and increased tax revenues, supporting schools and infrastructure.

Challenges and Solutions: Despite its potential, wind energy does face challenges. Wind generation is intermittent and site-dependent, which can lead to challenges in grid management and reliability. However, advancements in energy storage systems, such as batteries and other forms of energy storage, are mitigating these issues. Furthermore, there are concerns about wildlife impacts, particularly on birds and bats. This has led to better siting of wind farms and the development of technologies to detect and deter wildlife from turbines. Another challenge is public acceptance; the "not in my backyard" syndrome can be a significant barrier to development. Community engagement and shared economic benefits are critical in overcoming this challenge.

The Future of Wind Energy: The future of wind energy is vibrant and promising. With the advent of floating offshore wind turbines, the potential for wind energy expands even further, opening up new areas of the sea where winds are stronger and more consistent. Additionally, international commitments, such as the Paris Agreement, have underscored the need for renewable energy, with countries around the world investing in wind power. In a future that is oriented toward sustainability, wind energy is poised to play a pivotal role in the global energy mix, contributing to energy security, economic development, and environmental preservation.

Hydropower

Hydropower, harnessed from the energy of flowing water, is one of the oldest and most mature renewable energy technologies. From the water wheels of ancient civilizations to the complex hydroelectric power plants of today, the evolution of hydropower reflects humanity's enduring quest for sustainable and reliable sources of energy.

Principles of Hydropower: The fundamental principle behind hydropower is the conversion of the potential and kinetic energy of water into electrical energy. Water from rivers or stored in reservoirs at a higher elevation flows down through turbines, activating generators that produce electricity. The capacity for energy generation is dependent on both the volume of water flow and the height from which it falls, known as the "head." The predictability of water flow allows for a consistent and reliable generation of electricity, distinguishing hydropower from other, more intermittent renewable resources.

Technological Advancements: Technological advancements have significantly increased the efficiency of hydropower and reduced its environmental footprint. Modern hydro turbines are designed to maximize energy extraction and minimize ecological disruption. Innovations such as pumped-storage hydropower (PSH), which acts as a giant battery by storing energy in the form of water at an elevation, have enabled hydropower to offer grid stability and support for intermittent renewable sources like wind and solar. In addition, fish-friendly turbines and improved fish passage systems are being developed to mitigate the impact on aquatic ecosystems.

Benefits of Hydropower: Hydropower is lauded for a multitude of uses. Environmentally, it is a low-carbon energy source, emitting far fewer greenhouse gases than fossil fuels over its lifecycle. Economically, hydropower plants have long lifespans, often exceeding 50-100 years, providing long-term, low-cost energy and economic stability. Socially, large-scale hydropower projects can bring substantial benefits to local communities, including improved infrastructure, job creation, and flood control. Moreover, the global potential for small-scale, "run-of-the-river" systems can play a critical role in bringing electricity to remote, underserved regions without the need for large dams.

Challenges and Solutions: Despite its benefits, hydropower is not without its challenges. Ecological concerns such as habitat alteration, fish migration disruption, and changes in river sediment flow are significant issues. Addressing these challenges requires comprehensive environmental impact assessments and the incorporation of innovative design solutions to minimize negative impacts. Another challenge is the potential displacement of communities and the submergence of lands due to dam construction, necessitating careful planning and stakeholder engagement to ensure equitable outcomes. Climate change also poses a risk to hydropower's reliability, as altered precipitation patterns could affect water availability.

The Future of Hydropower: The future of hydropower is intrinsically linked to the global transition towards renewable energy. As efforts to decarbonize energy systems intensify, hydropower's role as a stabilizer for the grid and an enabler for other renewable sources becomes even more critical. The sector is likely to see an increase in retrofitting of existing dams with power plants, a rise in pumped-storage projects, and an emphasis on minimizing environmental and social impacts. With an eye toward sustainability, the hydropower industry is poised to continue its contribution to meeting global energy demands while supporting ecological and social objectives.

Geothermal Energy

Geothermal energy, derived from the Greek words "geo" meaning Earth and "therme" meaning heat, is a powerful and reliable renewable energy technology that exploits the vast reservoirs of heat stored beneath the Earth's surface. Unlike solar and wind energy, which are subject to the whims of weather and time, geothermal energy offers a consistent and ubiquitous energy supply.

Principles of Geothermal Energy: The Earth's core generates immense amounts of heat, remnants of the planet's formation, and radioactive decay processes. This geothermal energy manifests itself in visible features such as hot springs, geysers, and volcanic activity. Harnessing this energy involves tapping into the hot water and steam reservoirs within the Earth's crust to generate electricity or provide direct heating. There are various types of geothermal power plants, including dry steam, flash steam, and binary cycle plants, each designed to optimize the energy extraction process depending on the resource's temperature and pressure conditions.

Technological Advancements: Innovation in geothermal technology has significantly increased the viability and efficiency of this energy source. Enhanced Geothermal Systems (EGS), which involve injecting water into hot, dry rock to produce steam, have expanded the potential of geothermal energy beyond natural hydrothermal sites. Developments in drilling technology have also made it possible to access deeper and hotter geothermal resources, opening up new opportunities for energy extraction. Moreover, geothermal heat pumps (GHPs) utilize relatively stable temperatures just below the Earth's surface to heat and cool buildings, presenting a highly efficient alternative to traditional HVAC systems.

Benefits of Geothermal Energy: The benefits of geothermal energy are numerous. From an environmental perspective, it offers a low-carbon footprint, significantly less than conventional fossil fuels, helping to mitigate climate change. Geothermal power plants also occupy less land per megawatt than most other energy sources, preserving land for other uses. Economically, geothermal energy can provide a stable price over time, as it is largely immune to the volatile prices of fuel markets. The longevity and reliability of geothermal installations make them appealing for long-term investment and community planning. Furthermore, geothermal energy can spur economic development by creating jobs in plant operation, maintenance, and the broader supply chain.

Challenges and Solutions: Despite its potential, geothermal energy faces challenges. High upfront costs for site exploration, drilling, and plant construction can be prohibitive. However, as technology advances and economies of scale are realized, these costs are decreasing. There is also a risk of induced seismicity (man-made earthquakes) with the development of EGS projects, which necessitates careful monitoring and regulatory frameworks. Another challenge is the potential for the depletion of geothermal reservoirs if not managed sustainably. To address this, resource management strategies and reservoir re-injection techniques are being developed and refined.

The Future of Geothermal Energy: The future of geothermal energy is promising and could become an increasingly substantial part of the global energy mix. There is a growing recognition of its role in providing not only baseload power but also essential grid stabilization services in an era of increasing penetration of intermittent renewable sources. The global market for geothermal energy continues to grow, especially in regions with significant untapped geothermal resources, such as the Pacific Rim, East Africa, and parts of Europe. As the push for renewable energy intensifies, geothermal energy is set to expand, supported by policy incentives, technological advancements, and a growing commitment to sustainable energy.

Bioenergy

In the sphere of renewable energy, bioenergy emerges as a multifaceted technology utilizing biological materials, known as biomass, to produce electricity, heat, and fuel. As the quest for sustainable energy intensifies, bioenergy is seen as a vital component in the transition to a low- carbon economy. Unlike other renewable resources that are dependent on the elements, bioenergy's primary resource, biomass, is both storable and transportable, offering a unique reliability as a renewable energy source.

Operational Principles of Bioenergy: Bioenergy is derived through the conversion of organic matter, which includes plant materials, animal waste, and organic byproducts. The energy within this biomass is harnessed through biological processes like anaerobic digestion or thermal processes such as combustion, gasification, and pyrolysis. Anaerobic digestion releases biogas rich in methane, which can be burned to generate electricity or heat. Combustion directly converts biomass to heat and is a conventional method for electricity generation. Gasification and pyrolysis break down biomass at high temperatures in the absence of oxygen, producing syngas or bio-oil, respectively, which can then be processed into fuels or chemicals.

Technological Strides in Bioenergy: The technological landscape of bioenergy has witnessed transformative advancements. Second and third-generation biofuels are being developed from non-food biomass and microalgae, respectively, reducing competition with food supplies and improving the energy yield. Innovations in biochemical conversion processes have improved the efficiency of biofuel production, and thermal technologies like torrefaction have enhanced the energy density of biomass, making it more economical to transport and use.

Advantages of Bioenergy: Bioenergy's most compelling advantage is its potential to contribute to the circular economy. By utilizing agricultural residues, food waste, and sustainably sourced organic matter, bioenergy can reduce methane emissions from landfills and lower overall waste. It provides a pragmatic solution for rural and agricultural communities to convert waste into energy, fostering local development and job creation. Moreover, bioenergy systems can offer a carbon-neutral cycle, where the CO_2 released during energy production is offset by the CO_2 absorbed during the growth of the biomass, thereby maintaining a balanced carbon equation.

Challenges Facing Bioenergy: Despite its benefits, bioenergy is not without its challenges. There are concerns over land use, where the cultivation of energy crops may compete with food production or lead to biodiversity loss. Moreover, if not managed sustainably, the extraction and use of biomass can still lead to greenhouse gas emissions. Mitigating these concerns requires stringent sustainability criteria, efficient land management practices, and lifecycle analyses to ensure net positive environmental outcomes.

The Future of Bioenergy: Looking forward, the potential for bioenergy is substantial. With proper policy support and technological innovation, bioenergy can become more efficient and less resource-intensive. The integration of bioenergy with other renewable technologies, such as solar and wind, could lead to the creation of hybrid systems capable of providing a steady energy output. Furthermore, advancements in carbon capture and sequestration technologies could propel bioenergy to the forefront of the fight against climate change, transforming it into a carbon-negative energy source.

Ocean Energy

Ocean energy stands as a testament to human ingenuity in the quest for sustainable energy solutions. The world's oceans, covering more than 70% of the Earth's surface, are a vast, untapped reservoir of renewable energy. From the rhythmic dance of the tides to the relentless march of waves and the temperature gradients within the waters, the ocean is a perpetual motion machine that offers an array of energy-harnessing opportunities.

Various Forms of Ocean Energy: Ocean energy encompasses several forms, each with its unique method of generation and technology. The most prominent types include tidal energy, wave energy, ocean thermal energy conversion (OTEC), and salinity gradient energy.

(1) Tidal Energy: Tidal energy exploits the predictable rise and fall of ocean tides. Technologies such as tidal barrages and underwater tidal turbines can capture this energy, converting tidal movements into electricity.

(2) Wave Energy: Wave energy converts the surface motion of waves into

electrical power. Various devices, such as point absorbers, oscillating water columns, and attenuators, are designed to capture this dynamic energy.

(3) Ocean Thermal Energy Conversion (OTEC): OTEC utilizes the temperature differences between warmer surface waters and colder deep waters to drive a heat engine and produce electricity.

(4) Salinity Gradient Energy: Also known as blue energy, this form capitalizes on the energy potential from the difference in salt concentration between seawater and freshwater.

(5) Technological Advancements: Innovation in ocean energy technologies is on the rise. Advances in materials science have led to more durable and efficient designs for wave and tidal devices that can withstand harsh ocean conditions. Improvements in turbine technology enhance the conversion rates and reliability of tidal energy systems. In OTEC, new heat exchanger designs and working fluids are increasing system efficiencies. For salinity gradient energy, emerging techniques like pressure retarded osmosis and reverse electrodialysis are showing promise in laboratory settings.

Advantages of Ocean Energy: Ocean energy has several compelling advantages. It is abundant, with a vast potential to provide a significant portion of global energy needs. The predictability of tides and the consistency of ocean currents mean that, in certain aspects, ocean energy can be more reliable than other renewables, such as solar or wind. The ocean's proximity to many coastal cities can lead to reduced transmission losses and infrastructure costs. Additionally, ocean energy systems often have low visibility and minimal environmental footprint, especially when compared to land-based renewable energy systems.

Challenges: Nonetheless, ocean energy technologies face significant challenges. The high costs of development, deployment, and maintenance, driven by the need for robust materials and specialized ships, hinder widespread adoption. The marine environment poses technical challenges and risks, including biofouling, corrosion, and impacts on marine life. Regulatory frameworks and environmental impact assessments can also delay projects. Furthermore, grid integration and energy storage remain technical hurdles that need addressing for ocean energy to be scalable.

The Future of Ocean Energy: The future of ocean energy is closely tied to ongoing research and development aimed at overcoming current barriers. With sustained investment and supportive policies, ocean energy could play an instrumental role in achieving a diversified and resilient renewable energy mix. Cross-sector collaboration is also pivotal, with opportunities for sharing expertise from offshore oil and gas, maritime engineering, and environmental science to accelerate progress. In addition, as global concern for climate change mounts, ocean energy presents an opportunity for coastal nations to reduce their carbon footprint and harness local energy resources.

Hydrogen

Hydrogen, the lightest and most abundant element in the universe, holds the potential to revolutionize the energy landscape as a clean, renewable energy carrier. Unlike traditional energy sources that rely on carbon-based fuels, hydrogen offers a high energy yield and emits only water when consumed, presenting an appealing alternative to fossil fuels. As the world grapples with the urgent need to reduce greenhouse gas emissions and transition to sustainable energy systems, hydrogen energy emerges as a compelling path forward.

Production of Hydrogen Energy: The promise of hydrogen as a renewable energy resource lies in its versatility and the variety of methods available for its production. The most common method is electrolysis, where electricity, preferably sourced from renewable resources like wind or solar, is used to split water into hydrogen and oxygen. This process, when powered by renewables, is often referred to as "green hydrogen." Other methods include steam methane reforming and gasification of biomass, although these are not strictly renewable unless combined with carbon capture and storage technologies.

Storage and Distribution: One of the primary challenges with hydrogen energy is its storage and distribution. Hydrogen has a low volumetric energy density and must be stored under high pressure or at cryogenic temperatures. However, advances in materials science are leading to safer and more efficient hydrogen storage solutions, such as metal hydrides and advanced composite materials for high-pressure tanks. For distribution, existing natural gas pipelines may be repurposed for hydrogen, or new dedicated hydrogen pipelines could be constructed. Another approach is to convert hydrogen into more easily transportable compounds, such as ammonia, which can then be reconverted to hydrogen at the point of use.

Applications of Hydrogen Energy: Hydrogen's applications are diverse and encompass various sectors of the economy. In transportation, fuel cell vehicles powered by hydrogen have the potential to reduce emissions from the automotive industry significantly. Hydrogen is also a candidate for replacing fossil fuels in high-temperature industrial processes, such as steel and cement production. Furthermore, hydrogen can play a crucial role in energy systems as a way to store surplus electricity from intermittent renewable sources, acting as a buffer to balance supply and demand.

Challenges: Despite its potential, hydrogen energy faces several challenges. The production of green hydrogen is currently more expensive than hydrogen produced from fossil fuels, and the infrastructure for large-scale production and distribution is still in its infancy. There is also the challenge of public acceptance and the development of regulations and safety standards for handling and using hydrogen.

The Role of Hydrogen in the Global Energy Economy: Hydrogen has the potential to be a cornerstone of a zero-emissions future. It is particularly suited to complementing other renewable energy technologies by providing a solution for energy storage and sectors that are difficult to electrify. In the transition to a renewable energy economy, hydrogen can act as an energy carrier that helps to decouple energy production from consumption, both temporally and geographically.

12.2   ENERGY STORAGE TECHNOLOGIES

Energy storage technologies store energy for later use. These technologies are essential for balancing energy supply and demand, especially as renewable energy sources like wind and solar power become more common. Here are some common types of energy storage technologies:

Battery Storage

The global energy landscape is undergoing a significant transformation, pivoting away from fossil fuels towards renewable sources like wind and solar power. However, the intermittent nature of these energy sources necessitates a robust solution to store the generated power for use during periods of low generation or high demand. Battery storage systems have emerged as a pivotal energy storage technology, providing a critical solution to this challenge and unlocking the full potential of renewable energy.

Battery storage systems function by converting electrical energy into chemical energy for storage, which can then be reversed to supply electrical energy when needed. The significance of this process cannot be overstated, as it provides a buffer that can absorb energy during low-demand periods and discharge it during peaks. This capability is crucial for managing the intermittency of renewable energy sources like solar and wind, which do not produce electricity in a constant stream but rather in fits and starts, depending on the weather and time of day.

The Rise of Battery Storage Technology: Batteries have become the frontrunners in energy storage due to their scalability, declining costs, and improved technologies. The primary types of batteries used in energy storage are:

(1) Lithium-ion batteries are known for their high energy density and efficiency.

(2) Lead-acid batteries are traditionally used for their reliability and cost-effectiveness.

(3) Flow batteries are especially promising for large-scale storage due to their scalability and long discharge times.

Advantages of Battery Storage: Battery storage offers several advantages that make it an attractive option for energy storage:

(1) Modularity: Batteries can be easily scaled up or down depending on the need.

(2) Flexibility: They can be deployed anywhere, from residential to industrial scales.

(3) Rapid Response: Batteries can release stored energy almost instantaneously.

(4) Efficiency: Modern batteries have high round-trip efficiency, minimizing energy losses.

Integration with Renewable Energy Sources: Battery storage is instrumental in integrating renewable energy into the grid by:

(1) Smoothing out the variability of wind and solar power

(2) Storing surplus energy during peak production times

(3) Providing clean energy on demand, even when the sun isn't shining or the wind isn't blowing

Economic and Environmental Benefits: The deployment of battery storage has economic and environmental implications:

(1) Reduction in the need for peaker plants, which are often expensive and pollute more.

(2) Lower operational costs compared to conventional energy storage

methods.

(3) Decrease in greenhouse gas emissions by enabling a higher penetration of renewable energy.

Challenges and Future Prospects: Despite the progress, battery storage faces challenges that must be addressed:

(1) High initial investment costs, though declining, are still a barrier to widespread adoption.

(2) Concerns regarding the life cycle and sustainability of batteries, including the environmental impact of raw material extraction and end-of-life disposal.

(3) The need for technological advancements to improve capacity, durability, and safety.

The future of battery storage is promising, with research focusing on:

(1) New materials and chemistries to enhance performance and reduce costs.

(2) Improved recycling processes to create a circular economy for battery materials.

(3) Policies and incentives to accelerate adoption and integration into existing power systems.

Innovation and Future Prospects: The future of battery storage is bright, with ongoing research and development aimed at improving efficiency, capacity, and sustainability. Innovations such as solid-state batteries, which promise higher energy densities and improved safety profiles, are on the horizon. Efforts to create a circular economy for batteries through recycling and repurposing are also gaining traction. As such advancements materialize, battery storage is poised to play an even more central role in the transition to a cleaner, more resilient, and sustainable energy system.

Other Energy Storage Applications

These include:

(1) Pumped hydro storage facilities use excess electricity to pump water uphill. When electricity is needed, the water is released, turning turbines and generating electricity.

(2) Hydrogen can be used as a clean fuel and stored in fuel cells to generate electricity when needed.

(3) Compressed air energy storage that stores energy by compressing air and releasing it to drive a turbine generator.

(4) Flywheels store energy in the form of rotational kinetic energy.

(5) Thermal energy storage applications store energy by heating or cooling a medium (e.g., molten salt or ice) and then using that stored thermal energy to generate electricity when needed.

(6) Superconducting magnetic energy storage uses superconducting magnets to store energy in the form of magnetic fields.

12.3   CARBON CAPTURE AND STORAGE (CCS)

Carbon Capture and Storage (CCS) is a technology used to help mitigate the impact of fossil fuel emissions on global warming. It involves three main steps: capturing carbon dioxide (CO_2) produced from using fossil fuels in electricity generation and industrial processes, transporting the captured CO_2, and storing it underground in geological formations.

Capture

The first step in the CCS process is to capture the CO_2 produced by industrial processes and power generation. There are three main methods for capturing CO_2:

(1) Pre-combustion capture: In this method, fossil fuels are gasified to produce a mixture of hydrogen and carbon dioxide before combustion. The CO_2 is then separated, captured, and stored.

(2) Post-combustion capture: This method involves capturing CO_2 from the flue gas after combustion has occurred. The flue gas is passed through a solvent that absorbs the CO_2, which is then separated and captured.

(3) Oxy-fuel combustion: In this method, the fossil fuel is burned in oxygen instead of air, resulting in a flue gas that is mostly CO_2 and water vapor. The water vapor is then condensed, leaving nearly pure CO_2 that can be captured and stored.

Transport

Once the CO_2 has been captured, it must be transported to a storage site. This is typically done via pipelines but can also be transported by ship or truck in some cases.

Storage

The final step in the CCS process is to store the captured CO_2 underground in geological formations. These formations can include depleted oil and gas fields, deep saline aquifers, or deep coal seams. The CO_2 is injected into these formations, where it is trapped and stored permanently.

CCS is seen as an important technology for reducing greenhouse gas emissions, as it can capture up to 90% of the CO_2 emissions produced from using fossil fuels in electricity generation and industrial processes. However, there are also concerns about the cost and feasibility of implementing CCS on a large scale, as well as potential risks associated with underground storage of CO_2.

12.4   BIOENERGY WITH CCS (BECCS)

Bioenergy with Carbon Capture and Storage (BECCS) is a technology that combines bioenergy production processes, such as biomass combustion or fermentation, with carbon capture and storage (CCS) to remove carbon dioxide (CO_2) from the atmosphere. BECCS aims to generate energy while simultaneously capturing and storing CO_2, thereby resulting in negative carbon emissions. Here's how the BECCS process works in detail:

Bioenergy Production

Biomass (organic materials such as crops, wood, and waste materials) is harvested and transported to a bioenergy production facility. The biomass is then processed and converted into bioenergy through various methods, such as combustion, gasification, or fermentation. The bioenergy produced can be used for electricity generation, heat production, or as biofuel for transportation.

Carbon Capture: During the bioenergy production process, CO_2 is released as a byproduct. As previously described, the CO_2 is then captured using various technologies, such as pre-combustion capture, post-combustion capture, or oxyfuel combustion.

Carbon transport & storage: Once captured, the CO_2 is transported to a storage site, usually deep underground in geological formations such as depleted oil and gas fields or deep saline aquifers. The CO_2 is then injected into the geological formation, where it is stored permanently, preventing it from entering the atmosphere.

Key Benefits

Negative Emissions: The main benefit of BECCS is its potential to generate sustainable energy while creating a net negative emissions system. Since plants absorb CO_2 during their growth, and the CO_2 produced during energy generation is captured and stored, BECCS can result in negative net emissions. This means it removes more CO_2 from the atmosphere than it emits.

Flexible Energy Source: BECCS can also provide a renewable energy source, helping to reduce our reliance on fossil fuels.

Meeting Global Climate Targets: The technology can also contribute to meeting global climate targets, such as the Paris Agreement, which aims to limit global warming to well below 2 degrees Celsius above pre-industrial levels.

BECCS Challenges

Bioenergy with Carbon Capture and Storage is fraught with many challenges.

Land Use: Large-scale implementation of BECCS could require significant amounts of land to grow biomass. This can have substantial environmental and social impacts, such as deforestation and competition with food production.

Water Use: Biomass production can be water-intensive, leading to concerns about water scarcity in some regions.

Biodiversity Concerns: Monoculture plantations, if not managed responsibly, can reduce biodiversity and impact ecosystems.

Storage and Leakage Concerns: There are concerns about the long-term integrity of CO_2 storage sites and the potential for leakage back into the atmosphere.

Energy Penalty: Carbon capture processes can reduce the efficiency of power plants, as they require energy to operate.

Economic Factors: Current costs for BECCS can be high, and the economic viability depends on factors like carbon pricing, technology advancements, and policy support.

In conclusion, BECCS is a promising technology that has the potential to contribute to global climate mitigation efforts by creating a net negative emissions system. However, some significant challenges and limitations need to be addressed to ensure its sustainability and effectiveness in the long term.

12.5   ELECTRIFICATION OF TRANSPORTATION

Electrification of transportation is one of the most effective strategies for mitigating climate change. The transportation sector is a major contributor to greenhouse gas emissions, primarily because most vehicles rely on fossil fuels like gasoline and diesel. We can significantly reduce emissions and reliance on fossil fuels by switching to electric vehicles (EVs) and improving the electric grid's efficiency.

Here are some key points regarding the electrification of transportation to mitigate climate change:

Switching to Electric Vehicles: Electric vehicles (EVs), including battery electric vehicles (BEVs) and plug-in hybrid electric vehicles (PHEVs), produce zero emissions at the tailpipe. This is a significant improvement over internal combustion engine vehicles that emit carbon dioxide and other harmful pollutants. Governments around the world are offering incentives to encourage the adoption of EVs. These incentives may include tax credits, rebates, grants, and investments in EV infrastructure, such as charging stations.

Improving the Electric Grid: To maximize the climate benefits of EVs, it is essential to transition the electric grid to renewable energy sources such as solar, wind, and hydro. This will reduce the carbon footprint of charging EVs. Investments in smart grid technology can improve the electric grid's efficiency and facilitate the integration of renewable energy sources.

Developing EV Infrastructure: The availability of public and private charging stations is crucial in the widespread adoption of EVs. Governments and private companies are investing in the development of charging infrastructure to make it easier for people to charge their EVs. The development of fast-charging technology is also essential to reduce the time it takes to charge EVs, making them more convenient for long-distance travel.

Research and Development: Investing in research and development can lead to technological advancements that improve the efficiency and performance of EVs. This includes improving battery technology, which is a key factor in the range and cost of EVs. Governments and private companies are funding research to develop new materials and technologies that can improve performance and reduce the cost of EV batteries.

Public Transportation and Active Transport: Electrifying public transportation, such as buses and trains, can also contribute to reducing emissions from the transportation sector. Encouraging active transportation, such as walking and cycling, and investing in infrastructure that supports these modes of transport can further reduce emissions.

Policy and Regulation: Governments can play a crucial role in promoting the electrification of transportation through policies and regulations. This includes setting vehicle emissions standards, investing in public transit, and providing incentives for EV adoption. Some countries and cities also implement bans on internal combustion engine vehicles in certain areas or by specific dates, further promoting the shift to electric transportation.

Electrification of Transportation Challenges

The following are current challenges related to the electrification of transportation.

Technological: Battery technology needs further improvement in terms of energy density, cost, and charging speed.

Economic: High upfront costs of EVs, though this is decreasing as technology matures and production scales up. Infrastructure development requires significant investments.

Behavioral and Societal: Consumer acceptance and adaptation is a challenge. There is also the need for retraining and reskilling in industries affected by electrification.

In conclusion, the electrification of transportation is a crucial strategy for mitigating climate change. It requires a multifaceted approach that includes promoting the adoption of electric vehicles, improving the electric grid, developing infrastructure, investing in research and development, and implementing supportive policies and regulations. By addressing these factors, we can reduce emissions from the transportation sector and make significant progress in the fight against climate change.

12.6   SMART GRIDS

A smart grid is an advanced electricity supply network that uses digital communications technology to monitor and manage electric power's production, transmission, distribution, and consumption, thereby improving the electrical grid's efficiency, reliability, and sustainability.

Unlike traditional grids, smart grids allow two-way communication between the utility and consumers and can automatically adjust to changing electricity needs, improving efficiency and reliability.

The smart grid can play a crucial role in mitigating climate change by integrating renewable energy sources, reducing energy consumption, and improving the overall efficiency of the electrical grid. Here's a detailed description of how a smart grid can help mitigate climate change:

How Smart Grids Can Mitigate Climate Change

Integration of Renewable Energy: One of the chief advantages of smart grids is their ability to integrate renewable energy sources, such as wind, solar, and hydroelectric power, efficiently. By enabling better control and storage of intermittent renewable energies, they help reduce dependence on fossil fuels.

Demand Response: Smart grids allow for real-time adjustments to electricity demand. When supply is tight, utilities can offer incentives for customers to reduce consumption, thus avoiding the need to turn on peaking power plants, which are often less efficient and more polluting.

Energy Efficiency: The continuous monitoring and feedback loop in smart grids ensure that energy is used efficiently. For instance, if a line is experiencing losses, the system can identify it and rectify the problem.

Electric Vehicles (EV) Integration: Smart grids support the broad adoption of EVs by managing when and how they are charged, ensuring that the additional demand doesn't strain the grid and that renewable sources can be used as much as possible.

Energy Storage: Advanced energy storage systems can be integrated and optimized with the help of smart grids. Stored energy can be released when demand is high, reducing the need for fossil fuel generation.

Grid Reliability and Resilience: By rapidly detecting and responding to any disturbances, smart grids can prevent outages or quickly restore power. This means fewer blackouts and reduced need for excess capacity.

Microgrids: Smart grids support the development of microgrids - smaller, localized energy networks that can operate independently from the main grid. Local renewable sources can power these and can disconnect from the main grid during disturbances, ensuring a consistent power supply.

Smart Grid Challenges and Considerations

Cybersecurity: The increased connectivity and the two-way communication infrastructure introduce vulnerabilities. Protecting the grid from cyberattacks becomes paramount.

Interoperability: For a smart grid to function optimally, its various components, often from different manufacturers, must work seamlessly together.

Cost: Upgrading the current infrastructure to a smart grid requires significant investment.

Regulatory and Policy Framework: Policymakers need to formulate regulations that encourage the adoption of smart grid technologies and ensure their benefits reach all sections of society.

12.7   SMART CITIES AND INFRASTRUCTURE

Cities worldwide increasingly recognize the need to bolster their infrastructure to withstand extreme weather events and rising sea levels caused by climate change. Some of the key strategies and technologies being employed include:

Seawalls and Flood Barriers: Seawalls, levees, and flood barriers are designed to protect cities from storm surges and rising sea levels. These structures are typically made of concrete, steel, or Earth and are built along coastlines and riverbanks to prevent flooding. Examples include the Thames Barrier in London and the Maeslantkering in the Netherlands.

Stormwater Management Systems: Cities invest in advanced stormwater management systems to handle the increased rainfall and flooding associated with extreme weather events. This includes upgrading sewage systems, building retention basins, installing permeable pavement to allow water to seep into the ground rather than flooding streets, and large pump stations that can move vast quantities of water from city streets and canals. Examples include Tokyo's massive underground flood control system, the G-Cans project, and the pump stations in New Orleans, USA.

Green Infrastructure: Green infrastructure includes parks, green roofs, and other vegetated areas that can absorb rainwater and reduce flooding. These natural systems also help to cool cities and improve air quality. Examples include sponge cities in China and green roofs in Germany.

Building Design: New buildings in vulnerable areas are being designed to withstand extreme weather events, such as hurricanes and floods. This includes elevating structures, using flood-resistant materials, and incorporating features like rain gardens and bioswales to manage stormwater. Building codes may mandate wind-resistant designs, flood-proofing, or the use of certain materials. Examples include houses on stilts in many Southeast Asian countries and elevated roads in Miami.

Resilient Utility Infrastructure: Underground power lines, backup power sources, water storage, and communication redundancies to ensure continued access to power, water, and communication during and after extreme events.

Early Warning Systems: Cities are investing in advanced early warning systems that can provide residents with timely information about impending extreme weather events. These systems use satellite technology, sensors, and other tools to monitor weather patterns and predict when and where extreme weather events are likely to occur. An example is the Tsunami Warning Systems in the Pacific Ring of Fire nations.

Relocation and Managed Retreat: In cases where the risk is too great, cities might decide to relocate residents and critical infrastructure inland, essentially abandoning or repurposing vulnerable zones.

Climate Resilience Planning: Many cities are developing climate resilience plans that outline strategies for protecting infrastructure and residents from the impacts of climate change. These plans often include risk assessments, vulnerability analyses, and strategies for reducing greenhouse gas emissions.

Renewable Energy: Cities are investing in renewable energy sources, such as solar and wind, to reduce their reliance on fossil fuels and lower their greenhouse gas emissions.

Public Education and Engagement: Equip citizens with knowledge and resources to act before, during, and after extreme events. Cities do so with workshops, school programs, public service announcements, and community engagement initiatives.

Overall, the goal of these strategies is to make cities more resilient in the face of climate change by protecting critical infrastructure, reducing vulnerability to extreme weather events, and minimizing the impacts of rising sea levels.

12.8   CLIMATE-SMART AGRICULTURE (CSA)

Climate-smart agriculture (CSA) is an integrated approach to managing agricultural systems to ensure food security in the face of climate change. Its main goal is to assist farmers, agronomic advisors, and policymakers in securing sustainable and equitable increases in agricultural productivity and incomes while simultaneously adapting to climate change and reducing greenhouse gas (GHG) emissions where possible.

The following are the three main objectives of CSA:

Food Security and Increased Productivity: This ensures that food production systems are resilient enough to adapt to changes in climate and still meet the increasing demands of a growing population.

Adaptation and Resilience: Enhancing the resilience of agricultural systems to climate variability and change. This involves practices that reduce the susceptibility of farming systems to potential damages from climatic events and ensure their rapid recovery.

Mitigation: Reducing and/or removing greenhouse gas emissions where possible, making agricultural systems part of the solution to climate change rather than contributors to the problem.

It is a holistic approach that recognizes the interlinked challenges of food security, climate change adaptation, and mitigation.

Key Characteristics of CSA

Location-Specific: What may work as a CSA practice in one location might not be appropriate in another due to differences in local climate, soils, and socio-economic conditions.

System Approach: CSA considers all components of the agricultural system, including soil, water, crops, livestock, and the human components involved in production.

Incorporates Traditional Knowledge: It respects and utilizes the local and indigenous knowledge of farmers and communities, combining it with modern agricultural practices where appropriate.

CSA practices and technologies include

Soil Management: Techniques such as conservation tillage, agroforestry, cover cropping, and crop rotation help retain soil moisture, enhancing soil organic carbon and reducing soil erosion.

Water Management: Efficient irrigation systems, rainwater harvesting, and techniques such as alternate wetting and drying in rice cultivation help optimize water use.

Crop and Livestock Management: This includes the development and use of drought-resistant and early maturing crop varieties, improved livestock breeds, and integrating livestock with crop systems to recycle nutrients and energy.

Agroforestry and Silvopastoral Systems: Integrating trees with crops and livestock systems can provide multiple benefits like carbon sequestration, soil conservation, and income diversification.

Risk Management: This involves weather forecasting, crop insurance, and diversifying farm enterprises.

Energy Efficiency and Renewable Energy: Using biofuels, biogas, and solar/wind energy for farm operations helps reduce reliance on fossil fuels.

Post-harvest Management: Improved storage facilities and practices can reduce post-harvest losses, reducing wastage and the carbon footprint associated with wasted produce.

Supportive Policies and Financing

Governments and institutions need to provide supportive policies that facilitate the adoption of CSA practices. This includes research and development, capacity-building, and creating financial mechanisms such as grants, subsidies, or credit schemes tailored for CSA.

Stakeholder Involvement

Effective CSA needs a participatory approach. This means involving farmers, communities, NGOs, the private sector, and policymakers in decision-making and implementation.

In summary, Climate-Smart Agriculture represents a holistic approach to ensuring that our agricultural systems can weather the challenges presented by climate change while also being part of the solution. It combines traditional knowledge with modern technology, and it requires a supportive policy environment and active participation from multiple stakeholders.

13

HOPE, RESILIENCE, AND POSITIVE CHANGE

In the looming shadow of climate change, the future can appear fraught with despair and inevitability. The science is irrefutable: our planet is warming at an alarming rate, with consequences that ripple across ecosystems, economies, and societies. However, amidst the cacophony of doomsday predictions, there exists a powerful trilogy of human virtues: hope, resilience, and the capacity for positive change. These forces are the unsung heroes in the narrative of climate change, the subtle yet formidable tools that can drive humanity toward a sustainable and equitable future.

13.1 HOPE AS A CATALYST FOR ACTION

Hope is a powerful catalyst for action and can inspire individuals, communities, and nations to take meaningful steps toward mitigating climate change. Without hope, the enormity of the challenge might paralyze us into inaction. However, we are more likely to engage in meaningful efforts to mitigate and adapt to climate change when we have hope.

Hope can be derived from the innovative solutions and technological advancements that are emerging to combat climate change, as summarized in Chapter 12. Renewable energy sources such as wind, solar, and hydropower are becoming increasingly cost-effective and accessible, providing hope that a transition to a more sustainable energy system is possible. Additionally, reforestation initiatives and conservation efforts are helping to restore ecosystems and protect biodiversity, further instilling hope that we can mitigate the impacts of climate change.

The proliferation of global climate movements led by youth and indigenous activists is a testament to the enduring power of hope. These movements do not deny the gravity of the crisis; instead, they harness hope to mobilize action, advocating for policies and practices that mitigate harm and foster recovery.

Hope also fosters collective action. When communities, organizations, and governments share a vision of a better, more sustainable future, they are more inclined to work together to achieve common goals. Grassroots movements and international agreements, like the Paris Agreement, are a testament to the power of collective hope in addressing climate change.

At the individual level, hope can drive behavioral change. People who believe that their actions can make a difference are more likely to adopt sustainable practices, such as reducing energy consumption, conserving water, and embracing eco-friendly lifestyles.

13.2   RESILIENCE - ADAPTING TO A CHANGING CLIMATE

Resilience is the capacity to bounce back from adversity and adapt to changing circumstances. In the context of climate change, building resilience is crucial because we must prepare for and respond to the impacts that are already unfolding. Resilience takes various forms.

Ecosystems are inherently resilient and capable of adapting to gradual environmental changes. Resilience is found in the mangroves that shield coastlines from storm surges and in the drought-resistant crops that promise food security. Protecting and restoring ecosystems, such as wetlands and forests, can enhance our planet's ability to withstand the impacts of climate change.

Communities and individuals must also build resilience to cope with the effects of climate change. Resilience is woven into the fabric of communities that learn from each disaster, strengthening social ties and building knowledge that can be passed down through generations. This includes designing infrastructure to withstand extreme weather events, implementing disaster preparedness plans, and providing social safety nets for vulnerable populations.

13.3 POSITIVE CHANGE - INNOVATION & TRANSFORMATION

The marriage of hope and resilience begets innovation. The capacity for positive change is inherent to human nature. Throughout history, humans have faced monumental challenges and have responded with innovation, adaptability, and transformation. In the context of climate change, this capacity is more critical than ever. The urgent need for solutions has sparked a renaissance in climate technology and sustainable practices.

Advances in technology (see Chapter 12) have the potential to revolutionize our approach to climate change. Solar and wind power, electric vehicles, carbon capture and storage, and sustainable agriculture practices are just a few examples of innovations that can reduce greenhouse gas emissions and mitigate climate change.

Governments and international organizations have the power to enact policies and regulations that drive positive change. Carbon pricing, renewable energy incentives, and environmental regulations can all shape the behavior of industries and individuals toward more sustainable practices.

Shifting cultural norms and values can also drive positive change. As people become more aware of the environmental impact of their choices, there is a growing demand for sustainable products and services, which encourages businesses to adapt and innovate.

14

EDUCATION AND OUTREACH

As a multifaceted problem, climate change requires informed communities and individuals to drive tangible solutions. Education and outreach are critical tools in building climate awareness and mobilizing action to address the global climate crisis. By providing accurate, accessible, and relevant information, education and outreach efforts can empower individuals and communities to understand the scope and urgency of the climate crisis and take meaningful steps to mitigate its impacts.

14.1   CLIMATE LITERACY AND ACCURATE INFORMATION

Building climate literacy is crucial for enabling people to make informed decisions about their actions and advocate for effective climate policies. It includes understanding the science behind climate change and the social, economic, and environmental implications of the issue. Education and outreach efforts also help disseminate accurate and up-to-date information about climate science, the impacts of climate change, and potential solutions. This is essential for countering misinformation and assisting people to understand the severity and urgency of the issue.

14.2   INFORMING AND SHAPING PERCEPTIONS

Education lies at the heart of understanding the complex dynamics of climate change. Formal education systems have the power to shape perceptions from an early age, integrating climate science into curricula to foster a generation of environmentally literate citizens. Through structured learning, individuals are not merely informed about the scientific consensus on climate change but are also guided to comprehend the socio-economic and ethical dimensions of the crisis. Education instills critical thinking, enabling learners to navigate the misinformation that often impedes climate action. By embedding climate studies in disciplines such as geography, economics, and civics, education creates a comprehensive knowledge base that underscores the urgency of the crisis and the necessity for immediate action.

14.3 BRIDGING THE GAP BETWEEN KNOWLEDGE AND THE PUBLIC SPHERE

While education builds the foundation, outreach extends the reach of climate awareness into the public sphere, engaging diverse audiences beyond formal educational settings. Outreach programs facilitate dialogues between scientists, policymakers, and the public, ensuring that scientific insights translate into public understanding and policy. Initiatives like community seminars, public service announcements, and participatory workshops can demystify complex climate issues, making them relevant to everyday life. Outreach strategies often leverage media platforms to amplify their message, creating a ripple effect that can shift public opinion and increase pressure on decision-makers to prioritize climate action.

14.4 DRIVING BEHAVIORAL CHANGE

Effective climate education empowers individuals to make informed choices that reduce their carbon footprint. It promotes a feeling of control, equipping people with the tools to assess the environmental impact of their actions and to adopt more sustainable lifestyles. Engaging with young people is essential to climate education and outreach. Empowering the next generation with knowledge and skills to tackle climate change is crucial for ensuring long-term action on the issue. Young people can also become persuasive advocates for change within their own families and communities. Moreover, education fosters innovation by inspiring the next generation of scientists, engineers, and entrepreneurs to develop solutions to environmental challenges.

14.5   CATALYST FOR COMMUNITY ACTION

Education and outreach can inspire individuals and communities to take action to mitigate climate change, such as reducing greenhouse gas emissions, adopting sustainable practices, and supporting clean energy solutions. Tailored outreach efforts can unite local stakeholders around common environmental goals, fostering a collaborative approach to climate action. Community-driven projects like tree-planting campaigns, recycling drives, and sustainability fairs not only contribute to environmental conservation but also build social cohesion and resilience against climate impacts. Through these localized efforts, outreach transforms abstract global issues into tangible, actionable items that resonate with individual communities.

14.6   GLOBAL CONNECTIVITY AND SOLIDARITY

In our interconnected world, education and outreach must transcend borders to cultivate global citizenship. International educational exchanges and outreach programs can inspire solidarity and collective action, recognizing that climate change is a universal challenge that requires a unified response. Multilateral initiatives, such as the United Nations' Sustainable Development Goals, rely on global educational and outreach campaigns to galvanize support from all corners of the world.

15

FAMILY AND RELATIONSHIPS

The impact of family and relationships on climate change can best be illustrated by the example of our protagonist in the Personal Transformation story, Prof. Cole (see Chapter 1). It shows how the personal relationships of Prof. Cole, a climate denier turned believer, were affected by his transformation, including conflicts with friends or family members who may still be climate deniers.

15.1 FAMILY CONFLICT

The transformation of Prof. Cole from climate denial to environmental activism sent shockwaves through his family, creating an ideological chasm.

Prof. Cole's sister, Mary, is a strong climate change denier. In their first reunion at the refugee camp in Sierra Leone, Jamie was beginning to think that man's activities may be responsible for climate change. His sister did not feel that way. After Jamie returned to the USA from his trip to Africa, he kept in close contact with Mary and frequently discussed his shifting stance from denier to believer. Mary felt betrayed, and their relationship strained as they could no longer find common ground on the issue.

On the flip side, Jamie's son, Joseph, who is an environmental activist, was thrilled with his father's transformation. This strengthened their bond, and they collaborated on climate change awareness programs together. Jamie even invited Joseph to join him in the USA and his campaign for climate change action.

15.2 FRIENDSHIP TENSIONS

Friendships, too, are susceptible to the strains of such a transformation. Climate change, being a topic rife with political and social implications, often forms part of the ideological fabric that bonds friends together.

Some of Professor Cole's old friends from college, who shared his previous climate denial views, felt that he had 'sold out' or was seeking attention. He faced ridicule and resistance from friends who viewed the climate crisis through a lens of skepticism, causing a rift that was challenging to mend. They distanced themselves from him, causing a loss of long-standing friendships. Conversations that once flowed easily now stumble into uncomfortable silences or evolve into heated arguments, leading to an erosion of the mutual understanding that once existed.

Conversely, as Prof. Cole began attending climate change conferences and seminars, he made new friends who shared his newfound beliefs. These friends welcomed him with open arms, but the dynamics were still fragile as they sometimes questioned his genuine commitment to the cause.

15.3 PROFESSIONAL RELATIONSHIPS

Prof. Cole's colleagues at the university had mixed reactions. Some admired his courage to change his stance publicly, while others were skeptical and questioned his motives. This created a tense atmosphere at work, with Prof. Cole often having to prove himself. However, such a paradigm shift can also pave the way for integration into new environments that prioritize environmental awareness. Prof. Cole's transformation opened doors to new collaboration opportunities with other climate change researchers, expanding his professional network and giving him a chance to contribute positively to climate science.

16

ETHICAL AND MORAL DILEMMAS

Climate change presents one of the most complex ethical and moral dilemmas of the modern era. It is a global phenomenon with local impacts, and its causes and consequences are intertwined with issues of economics, social justice, and moral responsibility. The challenge lies not only in addressing the environmental changes but also in navigating the intricate human dimensions that these changes invoke.

16.1   THE PRECAUTIONARY PRINCIPLE

The precautionary principle holds that if an action or policy has the potential to cause harm to the public or the environment, in the absence of scientific consensus, the burden of proof falls on those advocating for the action or policy. The ethical dilemma here is how much evidence is needed to take action on climate change, given the uncertainty and complexity of the problem.

16.2   EXAMPLES OF MOST PRESSING DILEMMAS

Distribution of Responsibilities

One of the central ethical dilemmas of climate change is determining the distribution of responsibilities for both causing and addressing the problem. Historically, developed nations have contributed the most to greenhouse gas emissions, yet the developing world disproportionately feels the effects of climate change. This raises the question of fairness and whether those who have contributed least to the problem should bear the brunt of its consequences. The principle of "common but differentiated responsibilities" recognized in international agreements acknowledges this disparity, but translating it into effective policy remains contentious.

Another ethical dilemma relates to the responsibility of individuals versus collective entities (such as governments and corporations) to address climate change. With the knowledge that personal lifestyle choices contribute to global emissions, citizens in affluent nations face the moral question of reducing their carbon footprints. This raises further questions about the limits of individual responsibility in the face of systemic issues. While individual actions are essential, they may not be sufficient to address the scale of the problem. This raises questions about the role of government regulation and corporate accountability in mitigating climate change.

The transition to a low-carbon economy will have significant economic impacts, both positive and negative. There are ethical questions about how the costs and benefits of this transition should be distributed. For example, should workers in carbon-intensive industries be compensated for job losses? Should the costs of transitioning to renewable energy be borne by consumers or by the industries that have profited from fossil fuels?

Climate change threatens many species and ecosystems, raising questions about our ethical obligations to non-human animals and the natural world. Do we have a moral responsibility to protect species from extinction? How should we weigh the needs of humans against those of other species?

Climate change will likely force many people to leave their homes, either due to rising sea levels, desertification, or other impacts. What are the ethical obligations of nations to accept and help these climate refugees?

Uncertainties and Risks

While the science of climate change is robust, there are still uncertainties about the exact impacts and timelines. How do policymakers and individuals make ethical decisions facing these uncertainties? As sea levels rise and some areas become uninhabitable, we risk losing cultural heritage sites, and indigenous communities might lose their ancestral homes. How do we place a value on these losses, given the inherent uncertainties? Technological solutions, like carbon capture and solar radiation management, may help mitigate some effects of climate change. But they come with uncertain risks. Is it morally permissible to use these techniques, especially when the consequences could be global and irreversible?

Intergenerational Justice

Climate change is a profound challenge to the concept of intergenerational justice, which holds that we must leave the world in a condition that allows future generations to meet their own needs. The consequences of climate change, such as rising sea levels, extreme weather events, and changes in ecosystems, jeopardize the well-being of future inhabitants of the planet. The dilemma here is both moral and practical: what do we owe future generations, and how can we fulfill these obligations? The ethical frameworks that guide these decisions must balance present and future needs, but the unprecedented scale of potential future harm complicates these considerations.

Rights of Nature

The ethical discourse around climate change is increasingly recognizing the rights of nature. This paradigm shift entails seeing the natural world not merely as a resource for human exploitation but as an entity with intrinsic value and rights. The concept of granting legal personhood to rivers, forests, and ecosystems, as some cultures and legal systems have begun to do, poses a dilemma for conventional economic and legal structures that are primarily anthropocentric. This raises the question of how to reconcile human interests with the rights of the broader ecological community to which we belong.

Equitable Sharing of Resources and Adaptation

The dilemmas of climate justice also encompass the equitable sharing of resources and the opportunities for adaptation. As climate change threatens to exacerbate existing inequalities, the question arises as to who should have access to the limited resources and who should pay for the costs of adaptation and mitigation. Developing countries often lack the financial and technological means to adapt to climate change, and there is a moral imperative for wealthier nations to support these efforts. However, the tension between national interests and global solidarity creates a moral quandary for policymakers and citizens alike.

REFERENCES

1. Hansen, J; R, Reudy, M.Sato, and K. Lo, 2010: Global Surface Temperature change. Rev. Geophy., 48, RG4004. https://watershedtrust.ca/climate-change-update/ Accessed October 10, 2023

2. NASA 2023. Orbiting Carbon Observatory-2 https://ocov2.jpl.nasa.gov/ Accessed on October 25, 2023

3. NASA 2023. Orbiting Carbon Observatory-3 https://ocov3.jpl.nasa.gov/, Accessed on October 25, 2023

4. NOAA, 2023. Carbon dioxide, methane, and nitrous oxide rise further into uncharted levels. https://www.noaa.gov/news-release/greenhouse-gases-continued-to-increase-rapidly-in-2022. Accessed October 25, 2023

5. Statistica, 2023. Average carbon dioxide levels in the atmosphere worldwide from 1959 to 2022 https://www.statista.com/statistics/1091926/atmospheric-concentration-of-co2-historic/ Accessed October 10, 2023

LIST OF FIGURES